Detec

Defeating Distributed Denial of Service (DDoS) Attacks

BY

Seyed Mohammad Reza Khalifeh Soltanian and
Iraj Sadegh Amiri

Multimedia University, Malaysia
Photonics Research Centre, University of Malaya, 50603 Kuala
Lumpur, Malaysia

July 2014

ABSTRACT

Distributed Denial of Service (DDoS) attack is one of the most disruptive attacks in computer networks. It utilizes legitimate requests from hundreds or thousands of computers to specific targets to occupy targets' bandwidth and deplete targets' resource. In this work, we have attempted to not only mitigate DDoS attacks but also identify the source of attacks even behind Network Address Translation (NAT). This is followed by remedial actions such as denying further access or informing them that they have participated in the attacks.

This report presents a new algorithm to prevent servers from DDoS attacks. This algorithm requires that network routers or gateways collaborate with each other in order to detect suspicious traffic. The algorithm initiates a peer-to-peer communication among network routers or gateways to increase the probability of detecting unwanted traffic. We derive mathematical proofs based on cryptographic concepts such as birthday attacks to estimate the rate of attacks generated and passed along the routers. This implementation is to prevent the attacker from sending spam traffic to the server which can lead to DDoS attacks. The effectiveness of our implementation is evidenced in our experimental results.

TABLE OF CONTENTS

LIST OF TABLES

LIST OF FIGURES

CHAPTER 1

INTRODUCTION

1.1 Denial of Service and Distributed Denial of Service

Denials of Service (DoS) attacks are now one of the biggest issues in the Internet. It refers to malicious attempts to prevent legitimate users from accessing requested resources by depleting bandwidth or depleting the resource itself. Distributed Denial of Service is a large scale DoS attack which is distributed in the Internet. Every computer which has access to Internet can behave as an attacker. Typically bandwidth depletion can be categorized into flood and amplification attack. Flood attacks can be done by generating ICMP packets or UDP packets in which it can utilize stationary or random variable port. Smurf and Fraggle attacks are used for amplification attack (Specht & Lee, 2004). DDoS *Smurf attack* is an example of an amplification attack where the attacker sends packets to a network amplifier with the return address spoofed to the victim's IP address (Specht & Lee, 2004). Network amplifier is defined as a system capable of supporting broadcast addressing. Malformed packet and protocol exploit attacks are two most important ways to deplete resources. Nowadays, each of them can cause different types of application layer attacks (Yu, Lu, Fang, & Li, 2009). Application layer distributed denial of service attack is a DDoS attack that utilizes the communication protocol and sends out requests that they are indistinguishable from legitimate requests in the network layer. Moreover, most application layer protocols which are built on TCP communicate with users using sessions which includes one or many requests (Yu, Lu, Fang, & Li, 2009). HTTP1.0/1.1, FTP and SOAP are examples of application layer protocols. An application layer DDoS attack is categorized in one or a combination of the following types (Ranjan, Uysal, Swaminathan, & Knightly, 2006):

1. Rate of session connection requests sent by session flooding attack are higher than legitimate users.
2. Request flooding attack sends sessions that include more requests than normal.
3. Asymmetric attack sends sessions with more high-workload requests.

Constrained bandwidth and resource power are the major servers' weakness points and to overcome these issues we have to define specific threshold level for them as they are guaranteed quality of service (QoS). False rejection rate (FRR) is the fraction of the rejected requests over the total number of requests from legitimate users. Similarly, false acceptance rate (FAR) can be defined as the fraction of the accepted requests over the total number of requests from illegal users or attackers. Although a good DDoS defence mechanism has to reduce both FAR and FRR, reducing FRR has more privilege for the sake of user experience. Packet rate is used by most existing schemes as the metric to limit attackers (Li, Chang & Chan, 2005). Using packet rate as a metric to limit damage from attackers is the idea that if the source of the attacks can be identified and traced-back incrementally hop-by-hop to the source (or as close as possible), then rate limiting can be used to limit the scope and damage of the attacks. One of the famous traced-back packets to find the source of attacks is IP traceback scheme proposed by (Park & Lee, 2001) in which packets are randomly marked for tracking the routes of the attack packets. Rate limiting reduces the rate of the packets allowed through to routers or gateways. It is possible that intelligent attackers can adjust their packet rate based on server's response to evade detection. On the contrary, visit the clients' histories log are hard to be modified due to keep access logs by server, and the data used in trust evaluation are secured using cryptography. Hence using trust as the evaluation criteria will be more reliable in application layer DDoS attacks. One of the major properties of our solution to identify and mitigate

DDoS attacks which is distinct from other solutions is the manner in which routers and firewalls communicate to each other to reduce FRR and FAR as much possible as they can.

Distributed denial of service (DDoS) can take the major web sites down for several hours at a time by (Dean & Stubblefield, 2001). The attackers are able to break into hundreds or thousands of computers or machines and install their own tools to abuse them. Then they utilize these "zombie" machines to launch the DDoS attack as shown in Figure 1.1. Some of the tools even use encrypted communications between the attacker and zombie machines. To make the detection of the zombie machines hard as much as possible these tools counterfeit the source IP address on the traffic that they generate.

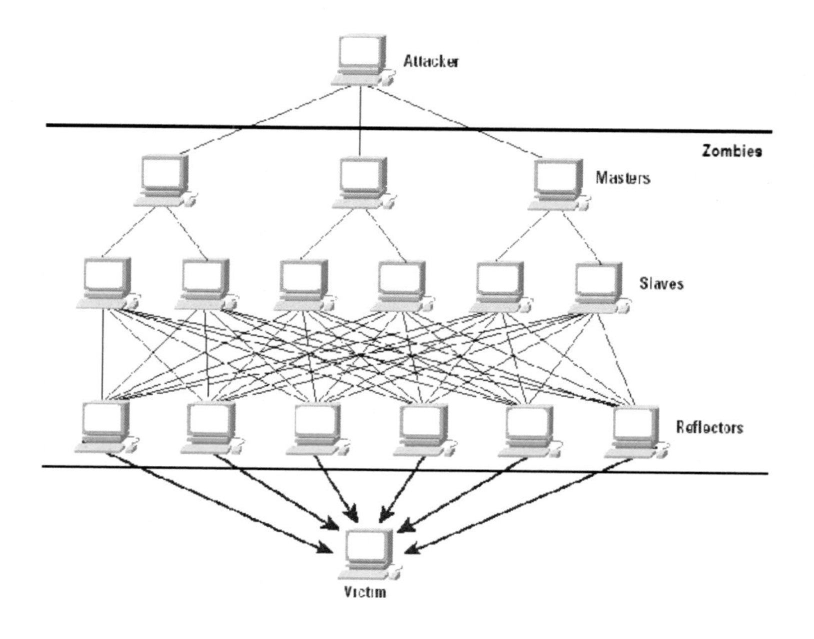

Figure 1.1. A single zombie network performs a DDoS attack.

The tools work via brute force which means random traffic is generated (may be with a political message) aimed at a specific machine. Most websites can bring down by generating gigabyte per second of traffic aimed at a single machine. For instance, the attacker could easily manage an attack such that the e-commerce sites remained available, but web surfers are unable to complete any purchases. Such an attack is possible by targeting the secure server that processes credit card payments instead. To make the clients' connection to such servers as secure as possible, the SSL/TLS protocol allows the client to request the server to perform a RSA decryption without having done any work first. One of the main disadvantages of RSA decryption is that it is an expensive operation. Only a few secure sites able to process 4000 RSA decryptions per second (Dean & Stubblefield, 2001). If we assume that a partial SSL handshake takes 200 bytes, then 800 KB/s is sufficient to paralyze an e-commerce site. Such a small amount of traffic is much easier to hide. Hence, this could be a major reason that DDoS attack can bring down even a very powerful server or router before taking any countermeasure. Up until now, there are lots of Internet service companies faced with DDoS attack. These attacks not only make some problems to the Internet users but also, impose severe financial losses to the targeted Internet companies that their business depends on online availability such as Amazon, ebay, Zappos, etc. Moreover, a DoS attack can initiate other malicious activities such as worm or virus infection and steal confidential information. DoS attack can be classified into two categories which are Bandwidth and Resource depletion. These categories are based on the inherent properties of the DoS attack which are either exhaust the resources of the server or attempt to deny the critical services. In Figure 1.2 the taxonomy of DoS or DDoS attack is illustrated.

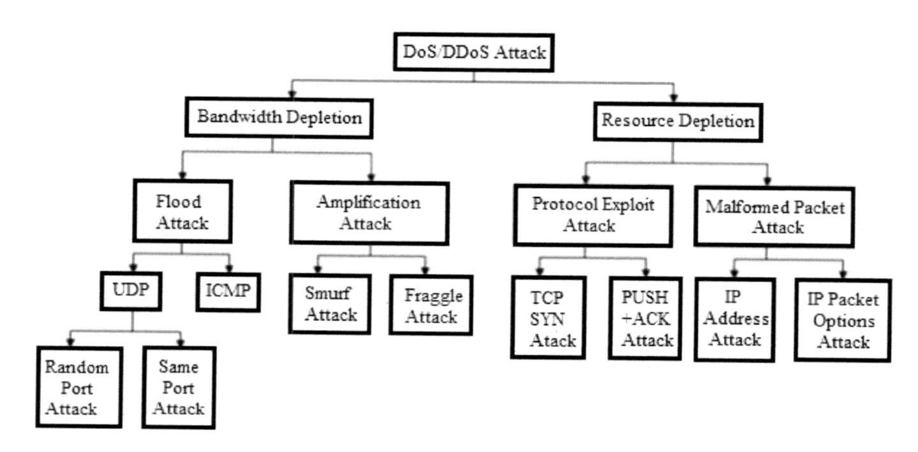

Figure 1.2. Taxonomy of DoS and DDoS (From Ting Kher Yee, 2011, p. 1)

1.2 Motivation

Sources of information via the Internet and demands for access to the Internet servers such as file servers, web servers and mail servers, increases exponentially. For instance, some malicious users known as spammers or spam DoS attackers abuse a popular server by sending spam request to overwhelm the server. Increasingly, spam requests are sent via zombie networks that work like a DDoS attack. Therefore, operations of the Internet service providers and also other Internet companies who have partnership or collaboration with those providers will be affected and experience service disruption. As a result, it would be crucial to mitigate the spam DDoS attack by having a secure deployable protection scheme applied on routers and server in the network.

1.3 Objective

The objective of this project is to propose a practical algorithm to allow routers to communicate and collaborate over the networks to detect and distinguish distributed denial of service attacks. This algorithm allows the detection of DDoS attacks on the servers as well as identifies and blocks the attacks.

1.4 Organization

The remaining parts of this report are organized into 4 chapters.

In Chapter 2, a survey the existing methods to identify and mitigate Distributed Denial of Services are presented. All related works are examined and discussed.

In Chapter 3, some preliminaries are covered which include the introduction of Mikrotik routers as infrastructural routers, concept of birthday attacks and birthday paradox. Also, we focus on the design and implementation of a comprehensive algorithm to identify and block source attacker. The work flow of the implementation is explained. Several real test cases, modelled to examine the efficiency of our theoretical and experimental implementation in defeating DDoS will be presented in the following chapter.

In Chapter 4, we present the results and discussions of our theoretical and experimental implementation. We also focus on obtaining an optimum parameter for router setting and make trade-offs to improve efficiency of mitigating DDoS attacks.

In Chapter 5, we conclude our achievement in this project based on whatever we found in our implementation. Last but not least, we make recommendations for future work to further improve the algorithms.

CHAPTER 2

OVERVIEW OF THE RELATED WORKS

2.1 General overview

Numerous attempts like the followings have been made as DDoS defence and response mechanism that include: packet filtering (Kim, Lau, Chuah, & Chao, 2004), IP traceback (Aljifri, 2003; Bellovin, Schiller, & Kaufman, 2003; Houle et al., 2001), flood pushback (Ioannidis & Bellovin, 2002), and client puzzle theory (Fraser, Kelly, Raines, Baldwin, & Mullins, 2007).

The capability to recognize the addresses of the true sources of the packets causing a DDoS is made available for the victim's network administrators by IP traceback methods that also comprise the introductory phase of recognition of the identity of any invader controlling a number of compromised machines in the DDoS. To enable victims for finding the attackers, there is continuous study in IP traceback, identifying the real source of a spoofed packet. In the following there are some problems with methods of IP traceback.

A slow complex process of logging into every router on the path may be regarded as one of manual ways. A number of costly routers might have processing abilities with internal software offering some stage of automatic IP traceback. A vigorous attack for the period of the testing of upstream links is required by link testing up to the time when the source is located. ISP co-operation and management overhead are necessary for input debugging. Numerous layers, usually comprising of blameless victims, exist in a Distributed Denial-of-Service network; the instigator may not even be active in the time of occurring the attack. Limiting the rate of ICMP and/ or SYN packets, verifying for correct reverse paths, using Ingress/Egress filtering (checking out packets , having valid inside source address, and having outside source destination), and preventing packet delivery from unidentified hosts may be conducted by firewalls. Sending info to other sources or saving info on the packet destination for future use, and marking them with additional info are regarded as Router Solutions through software enhancements.

Research in attack traffic filtering /packet filtering can be categorized into three areas according to the protection point:

2.1.1 Source initiated: Source sites are liable for assurance that there are attack-free outgoing packets. Examples comprise network ingress filters (Ferguson & Senie, 2000), disabling ICMP, or eliminating unexploited services to avoid computers from getting attack agents, or filtering odd traffic from the source (Mirkovic, Prier, & Reiher, 2002). The feasibility of these approaches, however, pivots on voluntary collaboration among a major number of ingress network administrators Internet-wide, making these methods somewhat unreasonable given the extent and the Internet uncontrollability.

2.1.2 Path-based: Only the packets pursuing the correct paths are permitted in this approach (Kuzmanovic & Knightly, 2003). Every packet with an incorrect source IP for a particular router port is assumed to be a spoofed packet and dropped. This eliminates up to 88% of the spoofed packets (Park & Lee, 2001). In another approach (Jin, Wang, & Shin, 2003) for a source IP, if there is wrong of number of travelled hops, the packet is dropped, eradicating up to 90% of the spoofed packets. These approaches are believed practical, but there is a somewhat high possibility of false negatives, i.e., falsely accepting attack packets. It seems that none of these approaches work when packets use unspoofed addresses, that are emerging trend.

2.1.3 Victim-initiated: Countermeasures to decrease received traffic can be started by the victim. For instance, in the pushback scheme (Ioannidis & Bellovin, 2002), the victim initiates decreasing excessive incoming traffic and asks for the upstream routers to conduct rate reduction, too. There exist other approaches according to packet marking (Kim, Jo, & Merat, 2003; Xu & Gue´rin, 2005), an overlay network (Keromytis, Misra, & Rubenstein, 2004), statistical processing (Kim, Lau, Chuah, & Chao, 2004; Li, Chang, & Chan, 2005), TCP flow filtering (Kim, Jo, Chao, & Merat, 2003; Yaar & Song, 2004), etc. Though victim-initiated protections are more favourable, some approaches are too costly to apply or need changes in Internet protocols.

On the other hand, traffic filtering method is costly, and confirmed to be susceptible in current attacks. Furthermore, ISPs are dependent on blocking of DDoS attacks and manual detection. If there is an attack, a subject-matter specialist executes an offline fine grain traffic analysis to recognize and distinguish the attack packets. Then, new filtering regulations on access control lists are created and fitted by hand on the routers. However, poor response time and fails to protect the victim are caused by the requirement for human interference before harsh harms are appreciated. In addition, the clarity of current rule-based filtering is too restricted, because it needs a clear specification of all kinds of packets to be removed.

As shown in Figure 2.1 and Figure 2.2, an overlay network is a computer network made on the top of another network. In the overlay, nodes can be suggested of as being attached by logical or practical links, so that each of them matches to a lane, possibly by many physical links, in the original network. For instance, distributed systems like peer-to-peer networks, client-server applications and cloud computing are overlay networks as their nodes run on top of the Internet. The Internet was basically made as an overlay upon the telephone network while through the advent of VoIP, today there is the telephone network changing into an overlay network made on top of the Internet more than before.

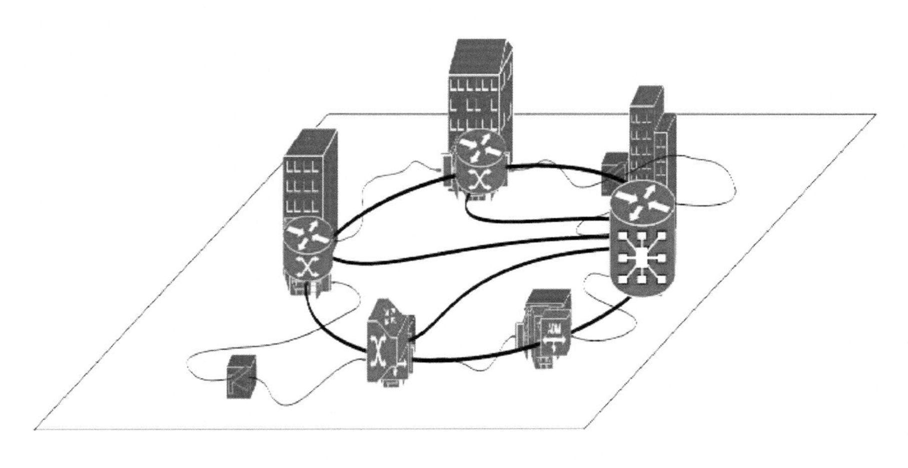

Figure 2.1. A sample overlay network

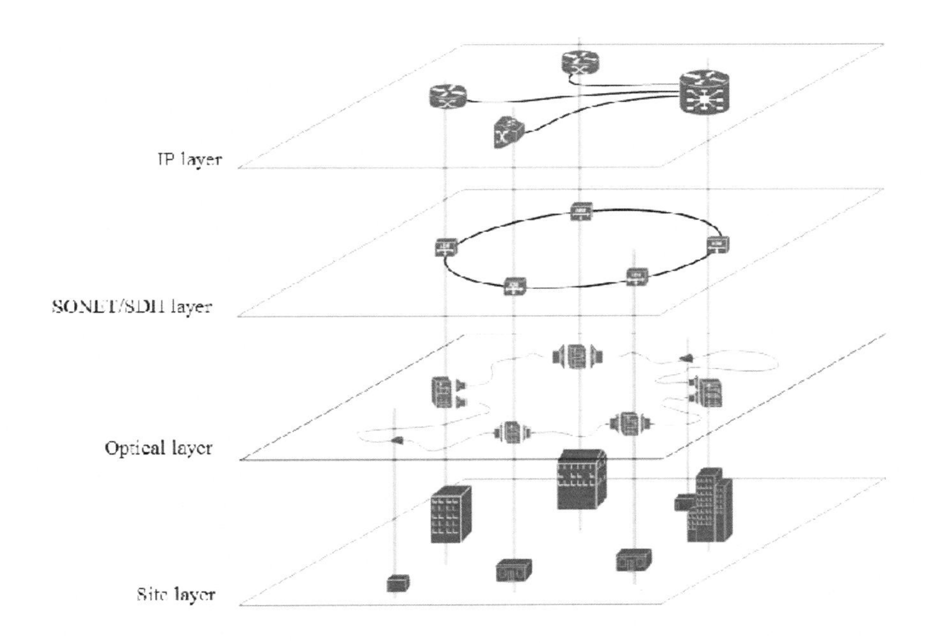

Figure 2.2. Logical layers of overlay network (From Wikipedia)

Overlay networks are applied in telecommunication networks due to the accessibility of digital circuit switching tools and optical fibres. Telecommunication transfer networks and IP networks joined to comprise the broader Internet are all overlaid with at least an IP or circuit layers if the public switched telephone network PSTN, an optical layer and a transport layer. At first enterprise private networks were overlaid on telecommunication networks like frame relay and Asynchronous Transfer Mode packet switching infrastructures. However, relocation from these infrastructures to IP based MPLS networks and virtual private networks began from 2001 to 2002 (AT&T History of Network Transmission).

In order to show Pushback, think about the network in Figure 2.3 in which the server *D* is under attack and the routers R_n are the last few routers through which traffic get to *D*. Links by which attack traffic is running are illustrated by the thick lines and the links with no bad traffic are shown through thin lines. Like the internal part of the network effectively conditioned, only the last link is actually congested. Hardly any non-attack traffic would be getting to the target in the lack of any particular measures. Some non-attack traffic is running by the links between R2-R5, R3-R6 R5-R8, R6-R8, and from R8 to D, but majority of it is dropped because of the R8-D congestion. *Bad* packets include which are transmitted by the attackers. An *attack signature*, which strived to identify characterizes bad traffic and the *congestion signature*, which is the group of characteristics of the aggregate identified as causing problems compromise something that can be really recognized. *Poor* traffic contains packets set the congestion signature, however, they are not actually part of an attack that are just unsuccessful enough to have the similar target, or some other characteristics that make them be identified as belonging to the attack. *Good* traffic is not equal to the congestion signature, but possesses common links with the bad traffic and may thus experience.

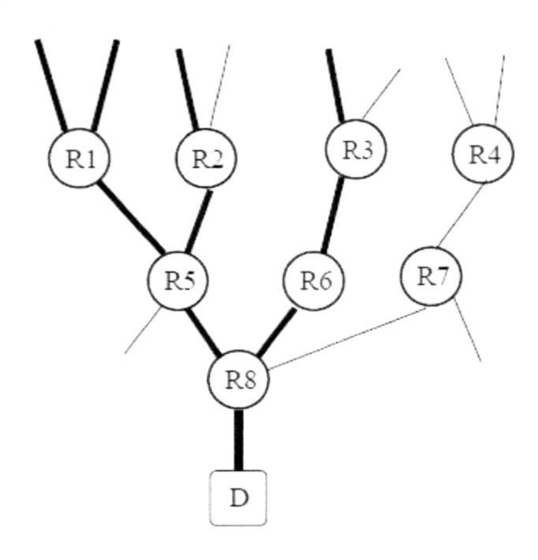

Figure 2.3. A DDoS attack in progress (From Ioannidis & Bellovin, 2002, p. 2)

Based on the figure above, the traffic entering R4 may be divided some good like the part exiting R7 that is not belonging to R8 and some poor, as belonging to *D*. From the links above, maybe there are some good traffic going to R5, and exiting from the lower left link, however, based on how congested the links R1-R5 and R2-R5 are, it may suffer. Further links include a combination of good and poor traffic. Then, it can do nothing to permit more good traffic initiating from the graph left side to reach *D,* regardless of how smart filters R8 could employ. If hopefully more good traffic would run in via R7, all it can perform is specially drop traffic from R5 and R6. By Pushback, R8 transmits messages to R5 and R6 notifying *them* to rate-limit traffic for *D*. Though the links downstream from R5 and R6 are not congested, packets are going to be dropped anyway when they arrive at R8, so perhaps they are dropped at R5 and R6 as well. These two routers, in sequence, broadcast the demand up to R1, R2, and R3, notifying *them* to rate-limit the bad traffic, permitting some 'poor', and more 'good' traffic, to run through.

According to extending a scheme of measured signature generation to incorporate anomaly detection system (ADS) with Snort by extracting signatures from anomalies noticed, more sophisticated intrusion detection systems (Hwang, Cai, Chen, & Qin, 2007; Ning, Jajodia, & Wang, 2001) have been recently suggested. In addition, DDoS defense schemes (Chen & Hwang, 2006; Kandula, Katabi, Jacob, & Berger, 2005; Moore, Voelker, & Savage, 2001; Walfish, Vutukuru, Balakrishnan, Karger, & Shenker, 2006) utilizing *Kill-Bots*, *Backscatter analysis* and experimental evaluation of *speak-up*, recently there have existed a defense against *application-level* DDoS.

In order to protect Web servers against DDoS attacks, *Kill-Bots* is a kernel extension that pretense as flash crowds. By providing authentication, Kill-Bots uses graphical tests that are different from other systems using graphical tests. First, to recognize the IP addresses that disregard the test, Kill-Bots uses an intermediate stage, and determinedly attack the server with demands despite frequent breakdowns at solving the tests. Since their intention is to obstruct the server, these machines are bots. On one occasion these machines are identified, Kill-Bots turns the graphical tests off, blocks their requests, and permits access to valid users who are incapable or reluctant to resolve graphical tests. Second, without permitting unauthenticated clients access to sockets, TCBs, and worker processes, Kill-Bots launches a test and controls the client's reply. The expected computing base (TCB) of a computer system is regarded as some of all hardware, firmware, and/or software parts that are serious to its security, in the sense that bugs or susceptibilities that occur in the TCB might endanger the security features of the whole system. Contrastingly, sections of a computer system out of the TCB must not be capable of misbehave in a way that it there would be any leak more than what is granted to them based on the security policy. Thus, Kill-Bots are expected to protect the authentication mechanism from being DDoSed. Third, Kill-Bots joins authentication with admittance control. A few restrictions are taken into account as the followings. First, Kill-Bots complicatedly deals with Web Proxies and NATs that multiplex a single IP address among several users. In case, all clients behind the proxy are legitimate users, then sharing the IP address has no impact. Contrastingly, when a zombie shares the proxy IP with legitimate clients and uses the proxy to mount an attack on the Web server, all subsequent requests from the proxy IP address may be blocked by Kill-Bots. Second, Kill-Bots has a few allocated parameters based on experience. Third, Kill-Bots supposes that the first data packet of the TCP link will include the *GET* and *Cookie* lines of the HTTP demand.

In an attack, *Backscatter analysis* supposed a per-packet random source address, consistent delivery and one reply generated for each packet. It states the possibility of a given host on the Internet obtaining at least one unsolicited reply from the victim is $\frac{m}{2^{32}}$ in an attack of m packets. Likewise, if one checks n distinct IP addresses, then the hope of occurring an attack would be:

$$E(x) = \frac{nm}{2^{32}}$$

Viewing a large enough address variety can "sample" such denial-of-service activity on the Internet. In these samples are information about the kind of attack, the victim's identity, and a timestamp from which can estimate attack period. Furthermore, considering these suppositions, can lead to use the average arrival rate of unsolicited responses directed at the monitored address range to approximate the real speed of the attack directed at the victim, as follows:

$$R \geq R' \frac{2^{32}}{n}$$

Supposing that R' is the backscatter calculated average inter-arrival rate from the victim and R is the extrapolated attack rate in packet-per-second, there are three main suppositions that underlie *Backscatter analysis*:

- *Address uniformity*: attackers spoof source addresses randomly.
- *Reliable delivery*: attack traffic is delivered consistently to the victim and backscatter is delivered consistently to the monitor.
- *Backscatter hypothesis*: unsolicited packets viewed by the monitor signify backscatter.

Amid these assumptions, key includes the random collection of source address. There exist some explanations why this supposition may not be legitimate. Firstly, *ingress filtering* (Fullmer & Romig, 2000) is employed by some ISPs on their routers to drop packets with source IP addresses outside the range of a customer's network. Consequently, an attacker's source address range might not comprise any of our intended addresses and we will undervalue the whole number of attacks.

"Reflector attacks" present a next problem for source address consistency. In this condition, by sending a packet spoofed with the victim's source address to a third party, an attacker "launders" the attack. And by sending a response back towards the victim, the third party responds. Third parties may further intensify the attack, when the packets to the third party are addressed applying a broadcast address (as with the popular smurf or fraggle attacks). The significant matter with reflector attacks is that the source address is particularly chosen. IP address is unable to view the attack, if it occurs in the range which is monitored as a reflector.

One more restriction is caused by *Backscatter analysis* supposition that packets are delivered consistently and that every packet produces a response. Probably throughout a great attack, packets from the attacker may be queued and dropped. Firewall or intrusion detection software may filter or may rate limit those packets that *do* arrive. Furthermore, some forms of attack traffic (e.g., TCP RST messages) do not normally extract a reply. Lastly, the responses may be queued and dropped themselves along the path back to monitored address range. As with random allocation supposition, these restrictions will lead to *underestimate* the number of attacks and their rate. However, they might also bias the classification of victims. The final restriction of *Backscatter analysis* technique is that it supposed unsolicited responses symbolize backscatter from an attack. There is freedom for any server on the Internet to transmit unsolicited packets to the observed addresses, and there may be misinterpretation of these packets as backscatter from an attack.

Speak-up, that is considered a defense against *application-level* distributed denial-of-service (DDoS), in which attackers cripple a server by sending legitimate-looking demands that use computational resources (e.g., CPU cycles, disk). Through speak-up, all clients and resources permitting are encouraged by a victimized server *to send higher quantities of traffic repeatedly*. It assumes that most of the attackers' upload bandwidth are already being used so that it cannot respond to the support. However, good clients have extra upload bandwidth so that they can respond to the support through severely higher volumes of traffic. The planned result of this traffic inflation is that the bad clients are crowded out by the good ones through which there would be capturing much larger fraction of the server's resources than before. *Speak-up* makes the server use resources on a group of clients in rough quantity to their collective upload bandwidths.

There have been many attempts by some of the researchers to defeat repeated DDoS attacks (Hussain, Heidemann, & Papadopoulos, 2006). In their endeavour, an approach has been performed to fingerprint and recognize the frequent attack scenarios. Such fingerprints are assumed to not only help in the prosecution of criminal and civil trial of attackers, but also assist in validating and focusing on response measures. As packet contents may be easily controlled based on the spectral description of the attack flow which are hard to falsify.

Some others apply trust-negotiation (Ryutov, Zhou, Neuman, Leithead, & Seamons, 2005) methods to establish trust, overlay networks (Wang, Chellappan, Boyer, & Xuan, 2006). Trust negotiation includes a method that offers an open verification and access-control surroundings for such transactions, but it is vulnerable to malicious attacks that result in denial of service or sensitive information leakage.

DDoS-resilient scheduling (Ranjan, Swaminathan, Uysal, & Knightly, 2006), D-WARD (Mirkovic & Reiher, 2005) and MultiLevel Tree for Online Packet Statistics (MULTOPS) (Gil & Poletto, 2001) were suggested for filtering and rate limiting on the flows supposed at the source-end. Often, Security managers tend to prefer to focus more on defending their own networks hence they selected home detection methods (Carl, Kesidis, Brooks, & Rai, 2006). COSSACK (Papadopoulos, Lindell, Mehringer, Hussain, & Govindan, 2003) and DefCOM (Mirkovic & Reiher, 2005) organize detectors at the victim side and transmit an alert to the filter or to the rate limiter that is placed at the side of the source. Chen and Song (Chen & Song, 2005) suggested a perimeter based scheme for Internet service providers to enable a service to defence DDoS attacks for their customers. They suggested a plan according to edge routers to realize the sources of the flood off attack traffics.

Change-point detection theory is used by most of the researchers to distinguish any unusual traffic distributed through the Internet due to DDoS attacks (Blazek et al., 2001; Chen & Hwang, 2006; Peng, Leckie, & Ramamohanarao, 2003; Wang, Zhang, & Shin, 2004). Because of the lack of precise statistics to explain about the pre-change and post-change traffic distributions, there would be a development of a nonparametric cumulative sum (CUSUM) scheme for its low computational intricacy (Blazek et al., 2001). The short-term behavior shifting from a long-term one is monitored by the scheme monitors. If the cumulative difference arrives at a specific threshold, there would be an attack alert. A central DDoS defense scheme was proposed by Wang et al. (2004) to check the change points of the gateway level. A similar approach was taken by Peng et al. (2003) in the source IP addresses monitoring.

Here, we are going to review two remarkable efforts which are used more recently and they are client puzzle theory and collaboration of detection of DDoS attacks over multiple networks.

2.2. Client puzzle theory

Much of the presented solution to overcome DDoS attacks includes enforcing servers to produce a puzzle when the servers feel under attacks and clients must resolve these puzzles and send it again to servers as shown in Figure 2.4. Verifying the solution by servers is the next step. Generally, framework applied to explain about any certain client puzzle's scheme includes main step proposed by Jeckmans (Jeckmans, 2009) and they are: *Setup*, *PuzzleGen*, *PuzzleSol*, and *PuzzleVer*. In general, they include the algorithms applied to open the requirements for puzzles formation, produce the client puzzles, solve the client puzzles and verify the solution of the client puzzle for the client puzzle system correspondingly.

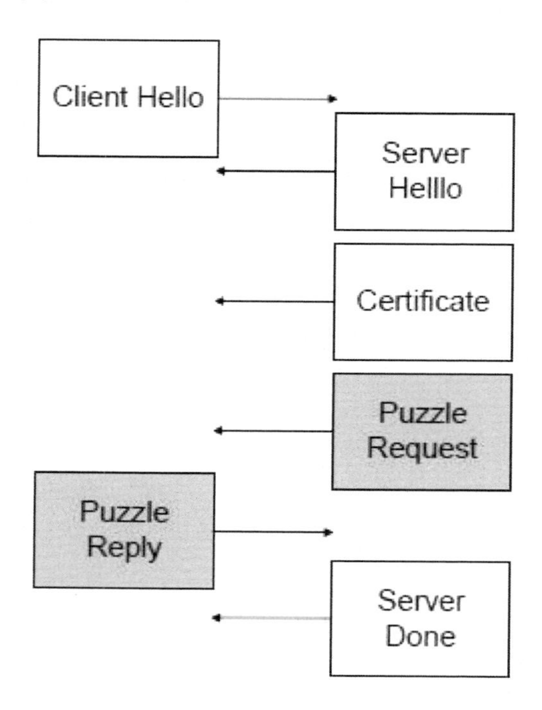

Figure 2.4. General Client puzzle handshaking (From Dean & Stubblefield, 2001)

Based on Figure 2.5 two main groups of, CPU-bound puzzles and memory-bound puzzles can be regarded as categorization of the existing client puzzles schemes. While memory-bound puzzles need some memory lookups to be performed to solve them, CPU-bound puzzles are the puzzles that require CPU cycles to be performed when solving them. Therefore, the solving speed would deeply rely on the processor and machine memory speed.

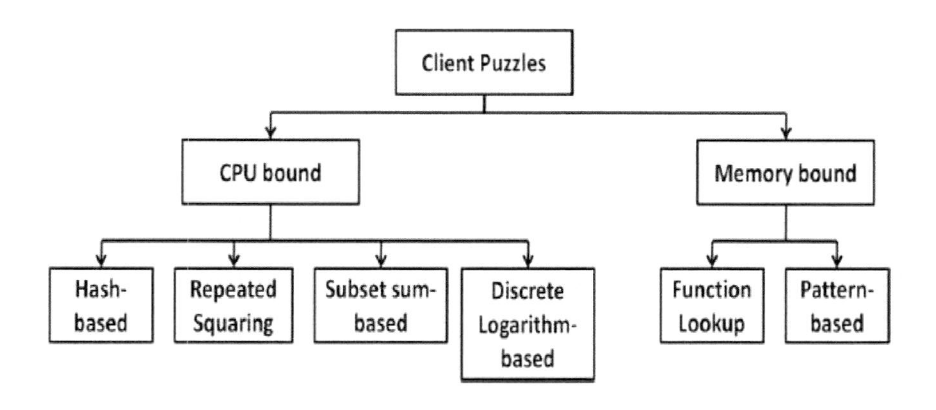

Figure 2.5. Classification of Client Puzzles Schemes (From Ting Kher Yee, 2011, p. 8)

Merkle (Merkle, 1978) suggested the first notion of cryptographic puzzles incorporated into network authentication protocol. Based on Merkle's idea, there are several client puzzles schemes which are applied to frustrate DoS resource depletion types of attacks (Gao *et al.*, 2010; Karame & Čapkun, 2010; Jeckmans, 2009; Dean & Stubblefield, 2001; Feng & Kaiser, 2007).

Several issues exist in performance of client puzzle which have to be carefully weighted up as the followings:

1. The price of puzzles' formation and answer confirmation used by the server has to be more costly than the price of solving the puzzle used by the client.
2. The complexity of puzzle ought to be modifiable.

3. There should be statelessness in the puzzle so that the server may not need to keep any link state prior to the offering of puzzle key.
4. For the clients to solve the puzzle, only limited amount of time is given.
5. There should be infeasibility in pre-computation attacks.
6. There is uniqueness in the puzzles, that is, by having the previous puzzles solved it is not possible to solve new puzzles.
7. There must not be any flooding attack vulnerabilities in puzzle creation and issuing.
8. Susceptibility to any bypass issue must not exist in puzzle mechanism.
9. There must be a differentiation in puzzle issuing between legitimate users and malicious users, therefore fine users exhibit spiteful behaviour in more complex puzzles.

Because of the application of client puzzles as a countermeasure against DoS attack, instead of developing a practical client puzzles system, most of the investigation has been centered on improving its security. There is any balance in the improvement of security and operation complexity of existing client puzzles schemes.

In general, client puzzle scheme not only cannot detect the attacker source, but also imposes both server and client heavy unnecessary computation. This is the inherent weakness property of client puzzle.

2.3. Collaboration of Detection over Multiple Networks

An algorithm was proposed by Y. Chen et al. (2007) in which a distributed method is proposed to identify DDoS flooding attacks at the traffic flow level. The defense system is appropriate for performance over the core networks operated by Internet service providers (ISPs). Some traffic fluctuations are detectable at Internet routers or at the gateways of edge networks at the early stage of a DDoS attack. There is a development of a *distributed change-point detection* (DCD) architecture applying *change aggregation trees* (CAT). The chief idea includes detecting abrupt traffic changes over multiple network domains at the initial time.

A new mechanism, called change aggregation tree (CAT) is applied by the algorithm that proposes a *distributed change point detection* (DCD) architecture.

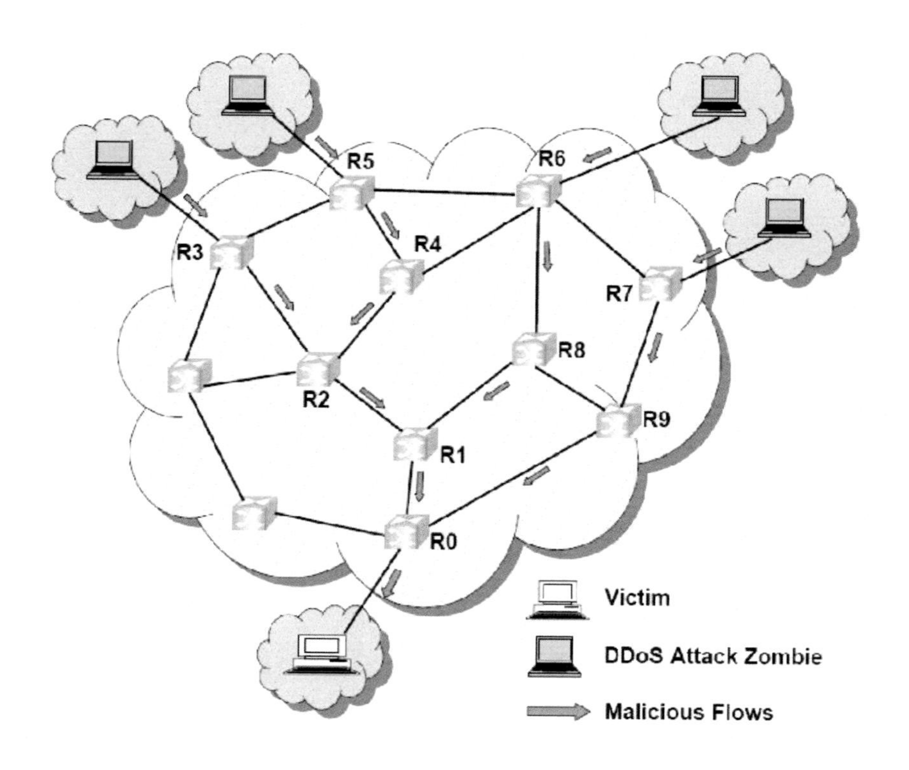

Figure 2.6. A large number of zombies generate traffic Superflow by DDoS flooding attacks toward a common destination as victim host (From Y. Chen et al., 2007).

The flooding traffic that is sufficiently large to crash the victim machine through communication buffer overflow, disk exhaustion, connection link saturation, and so forth. Figure 2.6 illustrates a flooding attack that is initiated from four zombies. The *attack-transit routers* (ATRs) identify the abnormal surge of traffic at their I/O ports. The attraction for the victim is by the end router *R0* in Figure 2.6. All the attack flows from the superflow homing toward the end router.

A Superflow includes all packets destined for the similar network domain from all possible source Internet Protocol (IP) addresses and uses a variety of protocols like Transmission Control Protocol (TCP) or User Datagram Protocol (UDP), etc.

Briefly, Y. Chen et al. (2007) developments may be classified in four technical features, as the quick vision and evidences of which are given in following segments:

1. *Traffic anomaly detection at the Superflow level.* Monitoring Internet traffic at routers on individual flows is identified by a 5-tuple: {source IP, destination IP, source port, destination port, protocol applied}. The Superflow includes those traffic flows destined for the same network domain and exploits the same protocol. For DDoS defense in real-life Internet environments, this level of traffic monitoring and anomaly detection is more cost-effective.

2. *Distributed change-point detection.* Bearing in mind the directionality of a DDoS flooding attack, in which the writers assume using joint routers for DCD and using the domain servers for alert correlation and aggregation.

3. *Hierarchical alerts and detection decision making.* The system accepts a hierarchical architecture at the domain and router levels. They suppose that their system

simplifies the alert relationship, global detection processes and allows the DCD system to implement in ISP networks.

4. *Novelty of SIP*. They assume as well a trust-negotiating SIP to secure interserver communications. The SIP has removed some of the shortcomings of the existing IP security (IPsec) and application-layer multicasting protocols (Kent & Atkinson, 1998; Wang, Chellappan, Boyer, & Xuan, 2006). SIP appeals for implementation on virtual private network (VPN) tunnels or over an overlay network built on top of all domain servers.

Through monitoring the propagation patterns of abrupt traffic changes at distributed network points, the DCD scheme is supposed to detect DDoS flooding attacks. An attack is asserted if a sufficiently large CAT is constructed to exceed a preset threshold. The principles behind the DCD system are presented by this behavior.

The system architecture of the DCD scheme is illustrated in Figure 2.7. The system is organized over multiple AS domains. In each domain there is a central CAT server. Traffic changes, aggregates suspicious alerts, checks flow propagation patterns and merges CAT subtrees from collaborative servers into a global CAT is detected by the system. The root of the global CAT is at the victim end. Each tree node and each tree edge correspond to an

ATR and to a link between the ATRs respectively. The abnormal surge of traffic at their I/O ports is detected by the *attack-transit routers* (ATRs).

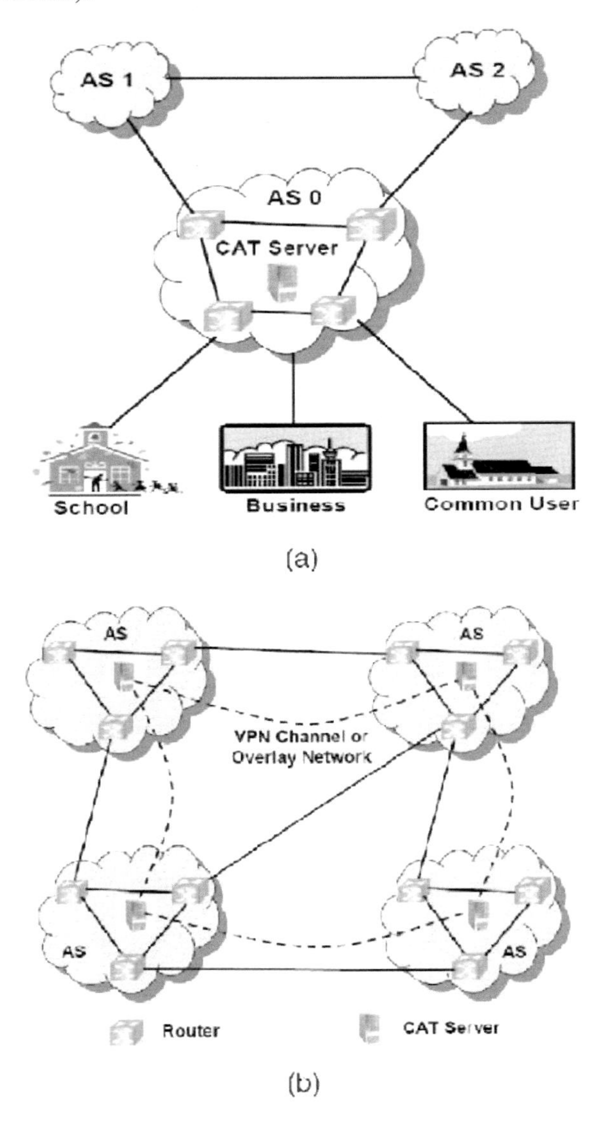

Figure 2.7. Distributed change detection of DDoS attacks over multiple AS domains. (a) Multidomain DDoS defense system. (b)

Interdomain communication via VPN tunnels or an overlay network atop the CAT servers in four domains (From Y. Chen et al., 2007).

CHAPTER 3

DESIGN AND IMPLEMENTATION

In this chapter, we introduce some tools which are utilized to design and implement of the algorithm. Also, we focus on the design and implementation of comprehensive algorithm to identify and block source attacker. The work flow of the implementation is explained and also the algorithm used in the implementation would be presented during this chapter. Moreover, several real test cases, modelled to examine the efficiency of our theoretical and experimental implementation in defeating DDoS will be presented in the following chapter.

3.1 MikroTik Routers

MikroTik is a Latvian company which was established in 1995 to develop wireless and routers ISP systems. Nowadays MikroTik provides software and hardware for Internet connectivity in most of the countries all over the world. They created the RouterOS software system in 1997 regarding to their experience in using industry standard PC hardware and complete routing systems which provides vast controls, flexibility, and stability for all kinds of routing and data interfaces.

Linux-based operating system known as MikroTik RouterOS is the main product of MikroTik. By installing the Mikrotik RouterOS on a personal computer or server computer, it turns the computer into a network router and enables implementing features such as virtual private network (VPN) server and client, firewall rules, wireless access point functions, bandwidth shaping, quality of service and other commonly used features for interconnecting networks and routing. Moreover, this system is able to serve as a captive-portal-based hotspot system. The operating system is licensed to promote service levels as far as each releasing provides more available RouterOS features. Winbox, a Windows software application, provides a graphical user interface for the RouterOS monitoring and configuration. Furthermore, the software allows connections via telnet, FTP, and secure shell (SSH). An Application Programming Interface also can be used for direct access from custom applications for monitoring and management.

One of the most outstanding features of Mikrotik routers is its ability to write our own program to make it more flexible. The packet generation feature is another tool that is able to generate and send raw packets utilizing random or constant port to evaluate performance of SUT (System Under Test) or DUT (Device Under Test). This traffic generator tool also collects jitter and latency values, counts lost packets, tx/rx rates and detects (OOO) Out-of-Order packets.

In general, majority of network packets are divided into three parts which are:

1. **Header:** The header carries information such as length of packet, synchronization, packet number, protocol, destination and source address.
2. **Payload:** This is also called the data or body of packet. It contains the actual data that the packet is delivering to its destination.
3. **Trailer:** It contains a few bits describing to the destination that the packet is finished. It also may contain some information for error checking.

We generate fake packets using Mikrotik to simulate the attackers (N_1 to N_L) in the network depicted in Figure 3.1. To generate this kind of packets we only manipulate the header of the packets and our algorithm does not care about of contents in payload and trailer of packets generated.

3.2 Multi Router Traffic Grapher (MRTG)

The Multi Router Traffic Grapher (MRTG) is a tool which allows administrators to monitor the traffic load pass through the network links. MRTG generates HTML pages containing PNG images that provide a LIVE visual representation of network traffic. MRTG uses SNMP which consists of a Perl script to read the traffic counters of routers and a fast C program to collect logs of the traffic data and creates beautiful graphs indicating the traffic on the network connection which has been monitored. These graphs can be viewed from any modern Web-browser.

MRTG creates visual representations of the traffic seen during a detailed daily view, the last seven days, the last five weeks and the last twelve months. This is possible regarding to the ability of MRTG to keep a log of all the data which has pulled from the router. MRTG is not limited to monitor traffic, it is also able to monitor any arbitrary SNMP variable. By using an external program we can even gather the data which should be monitored via MRTG. Usually, MRTG are using to monitor things such as

Login Sessions, Modem availability, System Load and more. MRTG even allows us to accumulate two or more data sources into a single graph.

3.3 Birthday Attack and birthday paradox

A birthday attack is a type of cryptographic attack which exploits the mathematics behind the birthday problem in probability theory. Birthday attack can be used in communication abusage between two or more parties. The attack depends on a fixed degree of permutations (pigeonholes) and the higher likelihood of collisions found between random attack attempts, as described in the birthday paradox/problem.

In probability theory, the birthday paradox or birthday problem considers the probability that, some pair people in a set of n randomly chosen of them, will have the same birthday. The mathematics behind this problem led to a well-known cryptographic attack called the birthday attack, which uses this probabilistic model to reduce the complexity of cracking a hash function (McKinney, 1966)

3.4 Legitimate requests and illegal requests

In this experiment, we modeled the legitimate user and the attacker (illegal user) with several attack strategies of different complexity.

3.4.1 Legitimate users

Legitimate users are the users who have the legal requests to the accepted destination from the edge routers' point of view (A_1, ..., A_k) and destination firewall routers (B) which are shown in the Figure 3.1. In the other words, legitimate users are end point entities who make connection requests for services provided by servers.

3.4.2 Illegal users or attackers

Attackers are end points whose requests are focused on depleting the target bandwidth or resources. In other word attackers try to prevent access of legitimate users to the particular server by keeping the number of simultaneous connection as large as possible. Sending session connection requests can be done with stationary rate or totally random.

The most important thing is examining and keeping the number of session connections which have the same destination address in each local network for specific timeslots under control. This plays an effective role to distinguish and mitigate DDoS attacks.

3.5 Traffic Models

To predict and find out how well a flow control scheme works we must modeled the behavior of network traffic. A developed model might involve higher-level protocols and characteristics of applications. It is necessary here to clarify and distinguish between "*smooth*" and "*bursty*" traffic.

Predictable and constant load is a result of smooth traffic source, or it might be achieved by changing only on time scales in which the time scales should be large compared to the response time of the flow control mechanism. It is easy to interact with this mentioned traffic. Regarding to share of the bottleneck bandwidth as fair as possible with little risk that some of the sources caused to underutilize links by stopping sending, the source can be assigned rates. Moreover, while bursts in traffic intensity are rare, switches can use a small percentage of available memory.

A good example for source of smooth traffic can be mentioned as fixed-rate compression video and voice traffics. Also, the effect of aggregating of a huge amount of bursty sources may be smooth, especially in a wide area network (WAN) where the loads from a huge amount of traffic streams are aggregated and the individual sources are uncorrelated and have relatively low bandwidth. On the other hand, Bursty traffic lacks any of the predictability of smooth traffic, as observed in some computer communications traffic (Leland, Taqqu, Willinger, & Wilson, 1993). Some types of bursts might be blocked by applications and users. For instance, a user expects to see a page or image so fast as possible by clicking on a link via World Wide Web browser. User's interactive response would be injured, since the network not only cannot predict when the clicks will trigger, but also should smooth out the resulting traffic.

Other sources of bursts result from network protocols that break up transfers into individual packets, remote procedure calls (RPCs), or windows, which are sent at irregular intervals. These bursts are not behave as steady state traffic due to their sporadic properties and typically do not have persistency long enough on a high-speed link to reach steady state over the link round-trip time.

In computer science, a remote procedure call (RPC) is an inter-process communication that allows a computer program lead to procedure or a subroutine to implement in another address space (typically on another computer on a shared network) without the programmer explicitly coding the details for this remote interaction. The programmer writes essentially the same code for both local and remote subroutine to the executing program. When the software in question uses object-oriented principles, RPC is called remote invocation or remote method invocation.

Designing control systems for smooth traffic is obviously much easier than designing flow control systems for bursty traffic. One of the most challenging parts which we face is design effective flow control for bursty traffic in order to support computer communications which are generally bursty.

3.6 Assumptions and considerations

Consider the network scheme as shown in Figure 3.1. In this scheme, we have assumed that there are k edge routers which have public IP address (ISP A_1 to A_k). We also have assumed that each of these networks has n subscriber in their private local networks. Therefore, there are $n \times k$ separate subscriber as we have also had in our internet network. Moreover, we have considered there are L end users that try to attack specific servers in the same subnet as network B. These users are distributed randomly in the network (N_1 to N_L).

Each of $n \times k$ end users is behind NAT. There are several servers such as SMTP, HTTP and FTP servers in the same subnet as network *B*, Subnet *B* is behind firewall. In our simulation router *B* communicate with each of edge router which recently have sent request to one of the *B* subnets. We also introduce some definition as shown in Table 3.1.

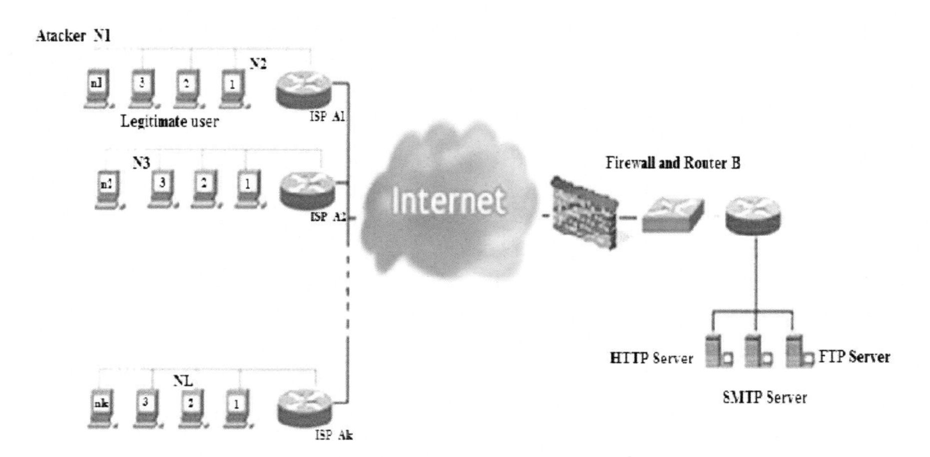

Figure 3.1. Simulated internet topology

Table 3.1. Table of definitions

$\overline{X}(t_m, i)$	Average number of packets or average input traffic load received by a router or firewall during time slot t_m at interface or port i
$\overline{Y}(t_m)$	Average CPU usage
$x(t_m, i)$	number of packets traffic load received by a router or firewall during time slot t_m at interface or port i
$y(t_m)$	Incident CPU usage at the time slot t_m
β_1	bandwidth threshold level
β_2	CPU usage threshold level
α	Inertia factor which has to satisfy $0 < \alpha < 1$ showing the sensitivity of the long-term average behavior to the current traffic variation
$S_{x,in}(t_m, i)$	deviation of input traffic from the average at time slot t_m
$S_{y,in}(t_m)$	deviation of incident CPU usage from the average at time slot t_m
$DFA_{x,in}(t_m, i)$	Dimensionless deviation of input bandwidth level from average (indicator of such an attack)
$DFA_{y,in}(t_m)$	Dimensionless deviation of CPU usage from average (indicator of such an attack)

3.7 Probability of concurrent website request

According to the recent survey which has been done by Netcraft, an Internet service company based in Bath England, there are around 366,848,493 on World Wide Web as of December 2011. It is important for us to know the probability of concurrent outgoing packets which have same destination address and different source address to set our algorithm threshold level based on that. It means that the algorithm which is used in edge routers has to determine threshold level for amount of outgoing packets per time slot. The amount higher than this threshold level indicates that the concurrency of some packets with the same destination address and different source address can behave as attack. Using the birthday attack problem allows us to determine this threshold level. Regarding to the generalized birthday problem, to a group of n people, where $p(n)$ is the probability of at least two of the n people sharing birthday, $p(n)$ would be:

$$p(n) = 1 - \overline{p}(n)$$

Where,

$$\overline{p}(n) = \frac{n! \cdot \binom{365}{n}}{365^n}$$

We can generalize this formula as *n* be the amount of internet users located in a same local area connection, for instance we had *n*=100 internet user in our experiment. The total number of different request possibility is the total number of active websites which has to replace with 365 in equation above. The probability of concurrency of at least 2 requests to the same destination in a short time slot is also depend on how much the requested website is well-known and famous. For example there must be difference between the amount of requests to the google.com and a non-famous personal website. By knowing the statistical average amount of this requests we have to define the time slot. In our experiment we define 60 seconds for this time slot as we knew the average requests to the router B. This means that we do not expect having more than 2 requests with the same destination address during 60 seconds. There is a trade-off to choose 60 seconds for this time slot. Any value less than 60 seconds gives more accurate results but imposes on the routers higher resource usage. For any values more than 60 seconds the probability of concurrency of 2 requests is not near zero. Also, this time slot depends strictly to *n*. If the edge routers find any two concurrent requests in the 60 seconds time slot, they can assume something is not normal and the edge routers are going to block the 2 concurrent requests if:

1. The edge routers received a packet from router *B* containing *"Flag"* tag.
2. The destination addresses of the 2 detected packets are *B*.

In our simulation in the real network which we will describe more about, we define 3 steps to become more accurate to detect the attacks. Each step would take 20 seconds to monitor all packets in the network.

3.8 Identifying and mitigating

In general, the mitigation and finding the source of DDoS attacks is a collaboration between edge routers ($A_1, \ldots, A_k$) and destination firewall routers *B* which has been shown in Figure 3.1. In this collaboration, router *B* will generate a packet under name of *"Flag"* and send to a group of edge routers one by one when it reaches its specific threshold which we describe more in detail later. Router *B* divides all of the edge routers IP address, which has sent request to its network, to *j* group of IP address in order to raise the speed of finding attackers and mitigation of DDoS attacks. When the edge routers receive this *"Flag"* packet, they run some scripts that we describe in the next section to find out whether or not the attackers are in their local networks. If they can find any source attacker then they reply the router *B* with a new packet under name of "*Response to Flag yes*" which means that we find attacker or attackers and add them to our block list to block their requests. If the edge routers cannot find any suspected source of attacks they would reply to the router B with a new packet under name of "*Response to Flag no*".

3.8.1 DDoS Detection algorithm on targeted server

In this part, we will describe how the router B can find out that the server is under DDoS attacks. As we describe about the properties of this kind of attacks, we can write scripts to make the router B sensitive to bandwidth depletion and resource depletion. Regarding to Table 3.1, average number of packets or average input traffic load received by a router during time slot m at port or interface i would be defined as (Y. Chen, K. Hwang, and Wei-Shinn Ku, 2007):

$$\overline{X}(t_m, i) = (1-\alpha)\cdot\overline{X}(t_{m-1}, i) + \alpha\cdot x(t_m, i)$$

And also the average CPU usage would be derived as:

$$\overline{Y}(t_m) = (1-\alpha)\cdot\overline{Y}(t_{m-1}) + \alpha\cdot\overline{Y}(t_m, i)$$

Where $0 < \alpha < 1$ is an inertia factor showing the susceptibility of the long-term average behavior to the current traffic variation. A higher α implies more dependence on the current variation. Y. Chen et al. (2007) define below $S_{x,in}(t_m, i)$ as the deviation of input traffic from the average at time slot t_m and $S_{y,in}(t_m)$ as the deviation of incident CPU usage from the average at time slot t_m (Y. Chen, K. Hwang, and Wei-Shinn Ku, 2007):

$$S_{x,in}(t_m, i) = max\{0, S_{x,in}(t_{m-1}, i) + x(t_m, i) - \overline{X}(t_m, i)\}$$

And similarly we can derive:

$$S_{y,in}(t_m) = max\{0, S_{y,in}(t_{m-1}) + y(t_m) - \overline{Y}(t_m)\}$$

The subscript *in* indicates that this is the statistics of the incoming traffic. While a DDoS flooding attack is launched, the cumulative deviation should be noticeably higher than the random fluctuations. Since $S_{x,in}(t_m , i)$ and $S_{y,in}(t_m)$ are sensitive to the changes in the average of the monitored traffic and resource usage respectively, the measurement of the unusual deviation from the historical average can be calculated as the equations below (Chen, Hwang, & Ku, 2007). The deviation from average (DFA) is the indicator of such an attack. The incoming traffic DFA is defined below at port i at time t_m (Chen, Hwang, & Ku, 2007):

$$DFA_{x,in}(t_m, i) = S_{x,in}(t_m, i) / \overline{X}(t_m, i)$$

And similarly the CPU usages as the resource depletion at time tm derive as:

$$DFA_{y,in}(t_m) = S_{y,in}(t_m) / \overline{Y}(t_m)$$

If the $DFA_{x,in}$ and $DFA_{y,in}$ exceed a router threshold β_1 and β_2 respectively, the measured traffic surge and resource over usage are considered a suspicious attack. The threshold β_1 and β_2 measure the magnitude of traffic surge and CPU usage over the average traffic and CPU usage value respectively. These parameters are preset based on previous router use experience. In a monitoring window of 100ms to 1 second, a normal Superflow is rather smooth due to statistical multiplexing of all independent flows heading for the same destination (Jiang & Dovrolis, 2005).

We expect a small deviation rate far below β_1 and β_2 if there is no DDoS attack. A Superflow contains all packets destined for the same network domain from all possible source Internet Protocol (IP) addresses and applies various protocols such as Transmission Control Protocol (TCP) or User Datagram Protocol (UDP), etc. In general, we assume the work range for parameter β_1 and β_2 in the range of $2 \leq \beta \leq 5$. In order to achieve best results we run our algorithm with two different values of router threshold setting β and compared results. Y. Chen et al. (2007) achieved an optimal router threshold setting $\beta \geq 3.5$ with an inertia ratio $\alpha = 0.1$. We choose β=3 with an inertia ratio $\alpha = 0.1$ and β=4 with an inertia ratio $\alpha = 0.1$ to make a trade off and find an appropriate router threshold setting.

Regarding to the definition of *β*, choosing any value below 2 makes the algorithm very sensitive to any small fluctuation of input traffic and CPU usage. Figure 3.2 shows the difference between choosing the β_1 value. In this graph the attack traffic is shown by the arrow. By setting any value below 3 for β_1, false-positive errors would be increased by detecting legal traffic request as attacks. On the other hand, by setting any value more than 5 for β_1, false-negative errors would be increased as the algorithm allows more attacks pass through the edge routers.

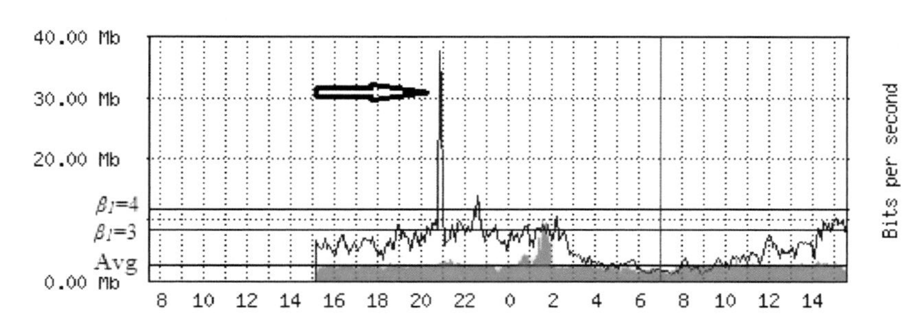

Figure 3.2. Effect of parameter β_1

When router *B* reaches the threshold level of β_1 and β_2, first the server put every 100 edge router's IP address in different *j* block or group. Next, the server generate a *Flag* packet to send to the first group which is $j = 1$ and wait for the response. After that, the router *B* has to wait due to running process in the edge routers for making any response. Once the waiting time has elapsed, the router *B* begins to receive response packet from the first group of edge routers which are either marked as "*Response to Flag yes*" or

"*Response to Flag no*". In general the whole process which is used in both targeted router B and edge routers A_1, ..., A_k is illustrated briefly as below:

1) Router B Calculate $\overline{X}(t_m, i)$, $\overline{Y}(t_m)$, $S_{x.in}(t_m, i)$, $S_{y.in}(t_m)$, $DFA_{x.in}(t_m, i)$, and $DFA_{y.in}(t_m)$
2) If $DFA_{x.in}(t_m, i) > \beta_1 \rightarrow$ go to 3 else go to 1
3) If $DFA_{y.in}(t_m) > \beta_2 \rightarrow$ go to 4 else go to 1
4) Send a *Flag* request from router B to a group or all of edge routers to ask them to check their networks for attacks.
5) Edge routers A_1, ..., A_k receive the *Flag* request and commence to monitor all outgoing packets.
6) If find at least 2 packet with the destination address of B and different source address in period of 60 second which is divided into three steps of 20 second $\rightarrow$ go to 7 else go to 8
7) Block or redirect this packets and send a reply packet to router B under the name of "*Respond to Flag Yes*" $\rightarrow$ go to 1
8) Send a reply packet to router B und/er the name of "*Response to Flag No*"
9) Go to 1

The flowchart in Figure 3.3 states the step by step sections of implementing scripts in the router B as the destination of DDoS attacks.

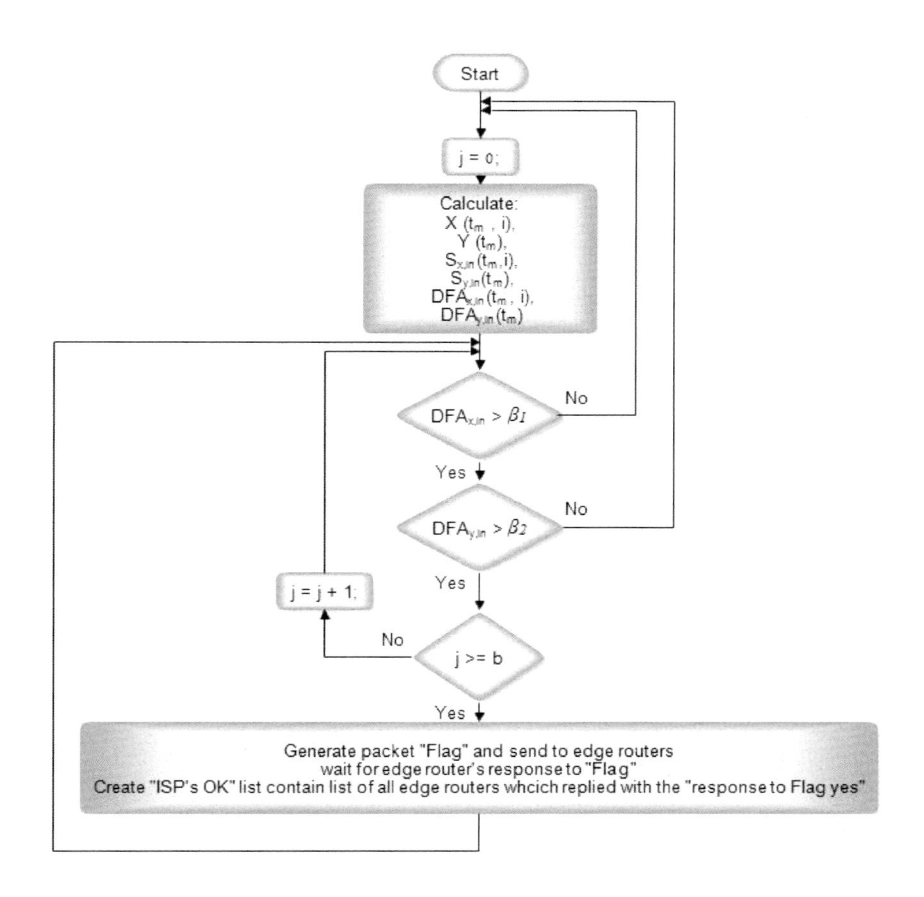

Figure 3.3. Router *B* as the destination of DDoS attacks' flowchart

3.8.2 Algorithm on Edge routers

As shown in Figure 3.1 Edge router is operate at the edge of an Multiprotocol Label Switching network that provides entry points into enterprise or service provider core networks. Edge routers also provide connections into carrier and service provider networks. We have to underline the importance of the role of these routers to help to identify and mitigate the DDoS attacks. In the other word, these routers are act as a gate of entrance for the local and public packets to the outside and inside a local area respectively. These gates also have access to the detail of packets which are flows through. With utilizing these gates we can identify and mitigate DDoS attacks.

The router *B* generates packets and sends to the destination address listed in group *j* and particular port of *x* which is known by routers A_1 to A_k. In this step router *B* is waiting for reply from each edge routers which are addressed in each block respectively. In edge routers we wrote a script as a scheduler to check if there is an incoming packet with destination port number of *x* to activate another script which is in charge of detecting attackers in its private network. The algorithm which is used in the Mikrotik firewall edge routers has been shown in Figure 3.4.

If the edge routers detect any attacks, reply to *B* with a packet which has destination port number of *x* as *'Response to Flag yes'* otherwise reply with destination port number of *y* as *'Response to Flag no'* which are known in the firewall router *B*. Router *B* get packet from all edge routers in the first block or some of them. If current received bandwidth and current CPU usage become below the threshold level then save the first block in the list of suspected source of attacks for its own records the same procedure is repeated to the second block of edge routers IP address list. These records can be also exchanged with other Firewall servers.

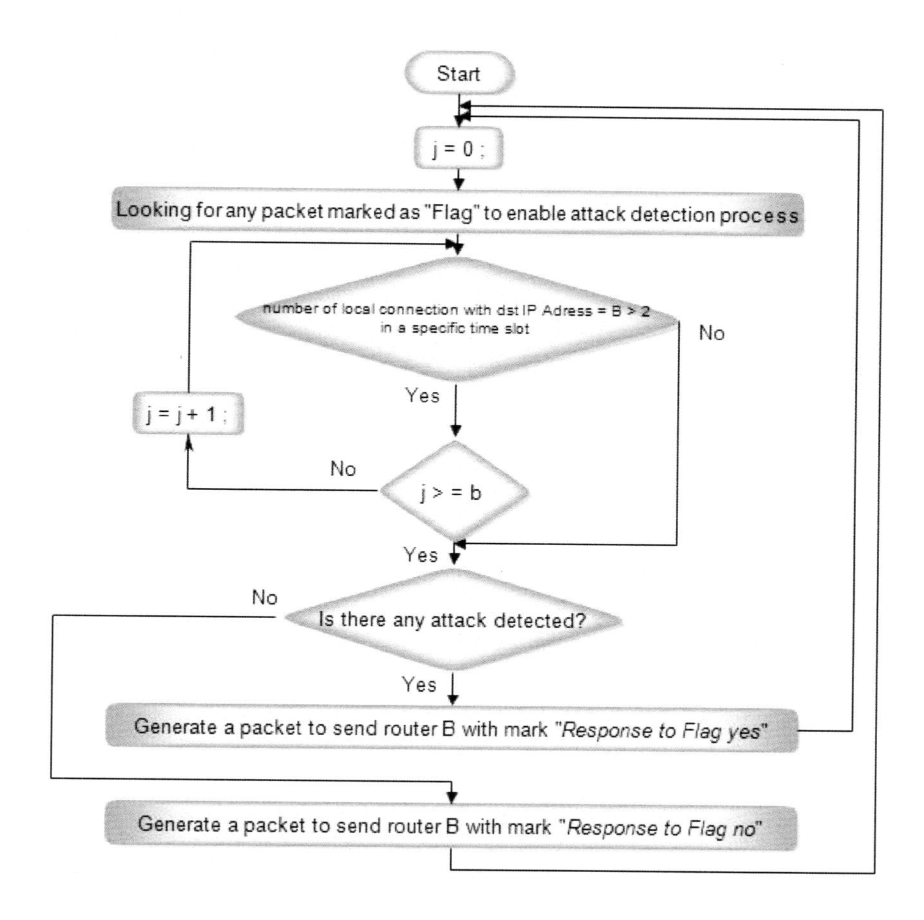

Figure 3.4. Flowchart of the edge routers scripts to distinguish attackers

CHAPTER 4

RESULTS AND DISCUSSION

In our topology that has been shown in Figure 3.1, we use Mikrotik routers as end clients to generate random UDP and TCP packets to destination *B*. Using Mikrotik Linux based firewall server gave us simulated behavior of attacks in large networks as well. The MRTG graph which is shown in Figure 4.2 shows the daily incoming and outgoing traffic of router *B* which takes 5 minute average. Figure 4.3 shows the daily CPU usage of 5 minute average under DDoS attack simulation and Figure 4.4 shows the daily CPU usage of 5 minute average without facing any attack. We generated random traffic to simulate DDoS attacks.

To generate random packets that behave like attackers, we use Mikrotik router. By using packet generator which is a Mikrotik tool, we generated fake traffic to the destination router *B*. The content of packet does not matter because the algorithm only analyses the packet headers. Source of packet, destination of packet, and number of packet sent by every individual user are three main factors that must be taken into consideration to behave as attack. After generating random traffic, we run the algorithm presented in Chapter 3 with router threshold setting of

- $\beta = 3$ with an inertia ratio $\alpha = 0.1$.
- $\beta = 4$ with an inertia ratio $\alpha = 0.1$.

We could detect simulated attacks and block the attackers very well. Figure 4.2 shows the generated attack from the edge router's LAN. The destination of these packets is router *B*. The edge router and router *B* refers to those seen in Figure 3.1. For further clarity, we show in Figure 4.1 the different interfaces on these routers.

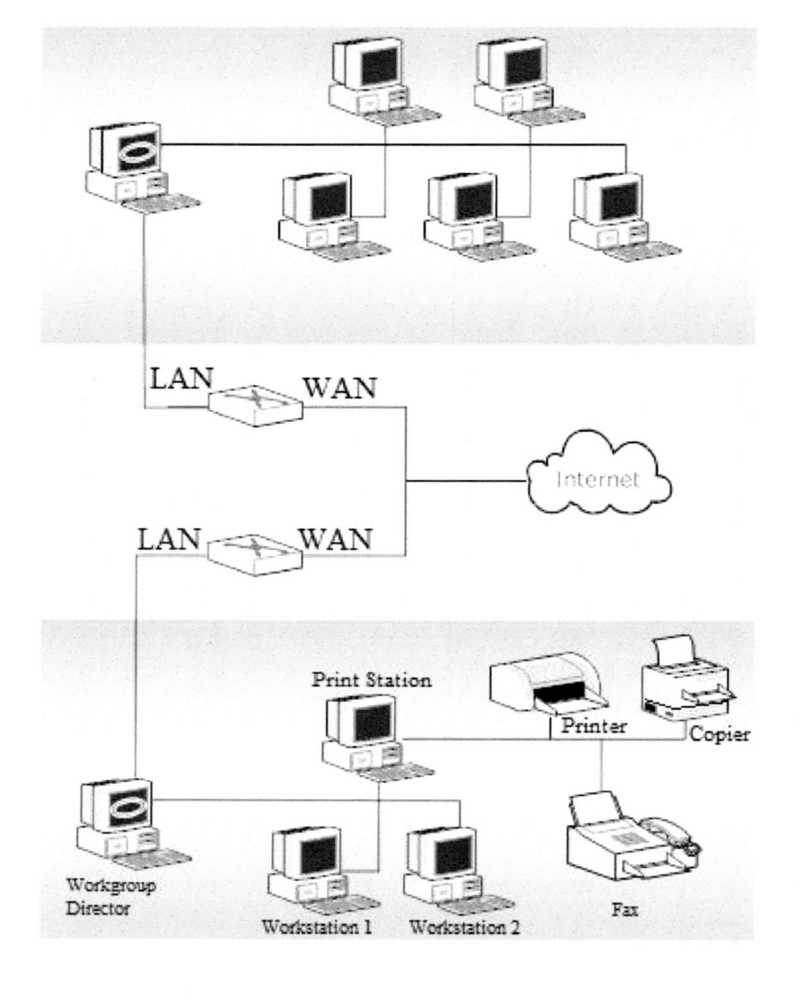

Figure 4.1. General WAN/LAN diagram

Edge routers consider the whole traffic as good traffic and route them to their destination address through its WAN interface that is shown in Figure 4.3. In other word, the edge routers do not care about where the packets are destined or whether the packets are potentially DDoS traffic. Figure 4.4 shows that the same amount of traffic that has been passed through the edge routers now act as attack traffic for router B LAN interface to deplete its bandwidth. This huge amount of request to the router B not only deplete a considerably percentage of available link bandwidth, but also can lead to deplete the resource in router B as shown in Figure 4.5. Figure 4.6 illustrate what happens when an attack is launched while running our algorithm. This figure shows the incoming generated attack packets in the LAN of edge router. For a short period of time edge router do not care about the traffic and let the packets go through. Router B finds out about the attack due to excessive use of the CPU and deviation of incoming average traffic from threshold level as shown in Figure 4.8 and Figure 4.9. In this time, router B send request to edge router as described in detail in Chapter 3 to enable scripts to commence the process of detecting and mitigating attack. After a while, edge router detects and mitigates the attack and blocks all sources of attack. The spikes in Figure 4.7, Figure 4.8, and Figure 4.9 which are shown with arrows indicate the short period of time that take to detect and mitigate the whole attack process by collaborating with the two most important elements of network that are edge routers and destined routers.

Time taken for the whole process is shown in Table 4.1 and Table 4.2 for ten independent experiments. Results for False-negative and False-positive error for different router setting for the two different parameters of β and α in 10 independent experiments are shown in Table 4.3 and Table 4.4.

"Daily" Graph (5 Minute Average)

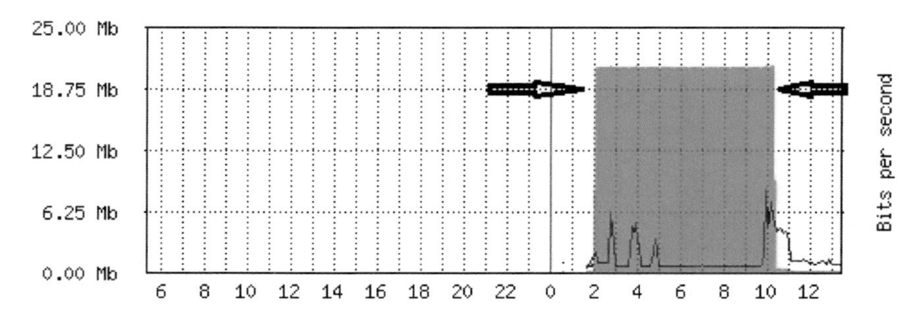

Max In: 21.19Mb; Average In: 14.27Mb; Current In: 37.07Kb;
Max Out: 8.51Mb; Average Out: 1.33Mb; Current Out: 763.04Kb;

Figure 4.2. MRTG daily incoming (green) traffic into LAN of edge router before launching our algorithm

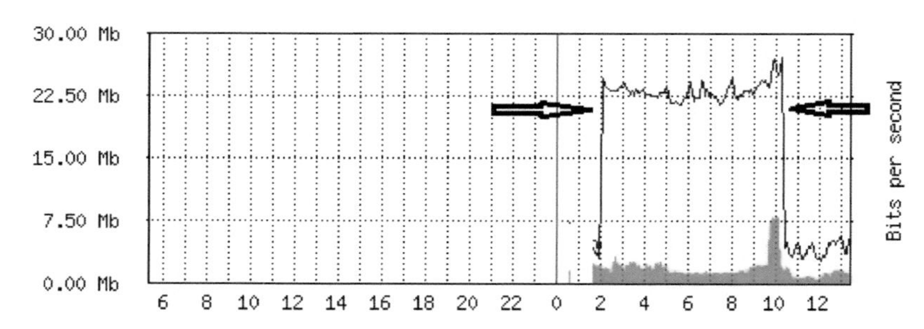

Max In: 7.89Mb; Average In: 1.59Mb; Current In: 1.15Mb;
Max Out: 27.35Mb; Average Out: 17.11Mb; Current Out: 5.18Mb;

Figure 4.3. MRTG daily outgoing (blue) traffic via WAN of edge router before launching our algorithm

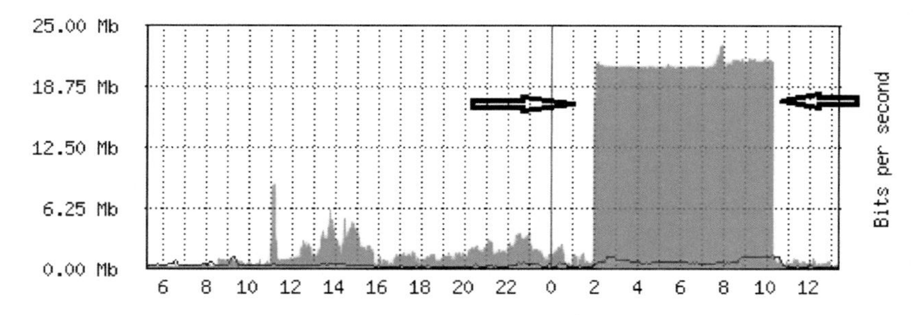

Figure 4.4. MRTG daily incoming (green) traffic into LAN of router *B* under attack before launching our algorithm

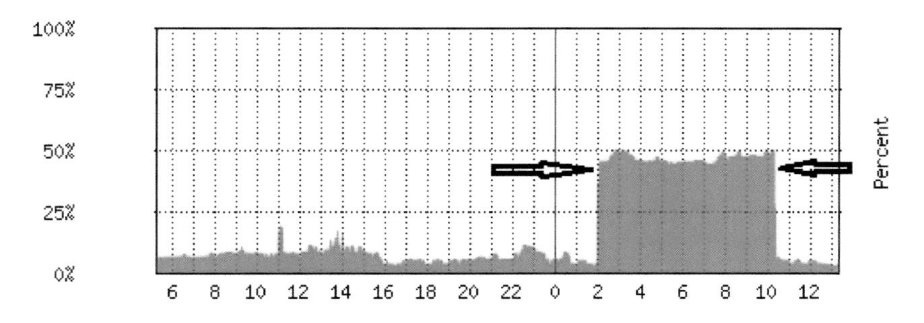

Figure 4.5. MRTG daily CPU usage of router *B* under attack before launching our algorithm

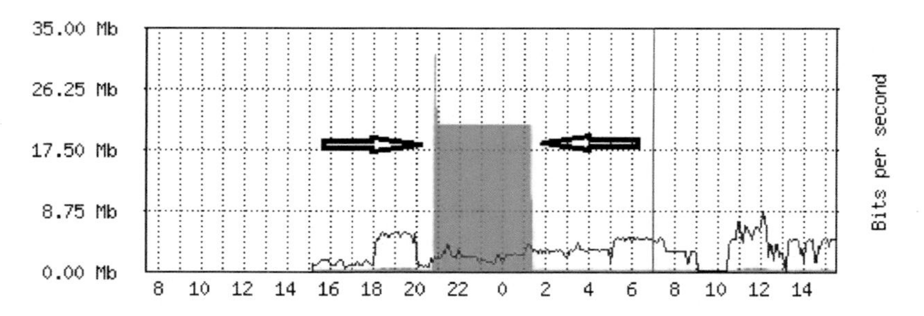

Figure 4.6. MRTG daily incoming (green line) traffic into LAN of edge router during running our algorithm

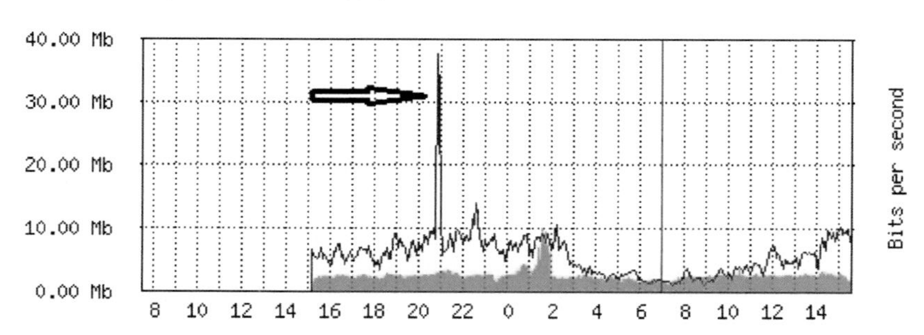

Figure 4.7. MRTG daily outgoing (blue line) traffic via WAN of edge router during running our algorithm

"Daily" Graph (5 Minute Average)

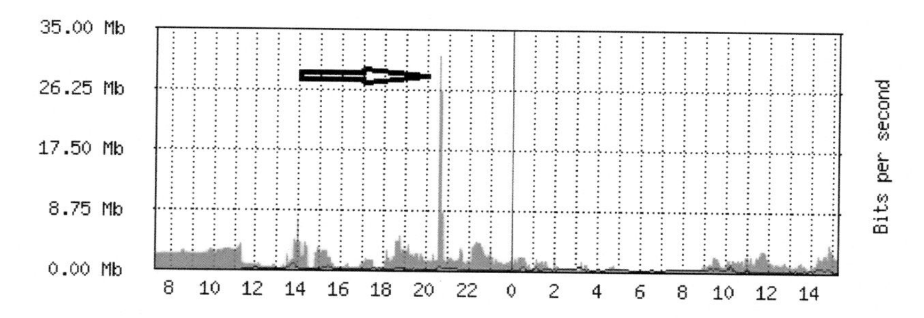

Figure 4.8. MRTG daily incoming (green line) traffic into LAN of router *B* during running our algorithm

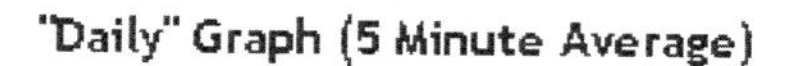

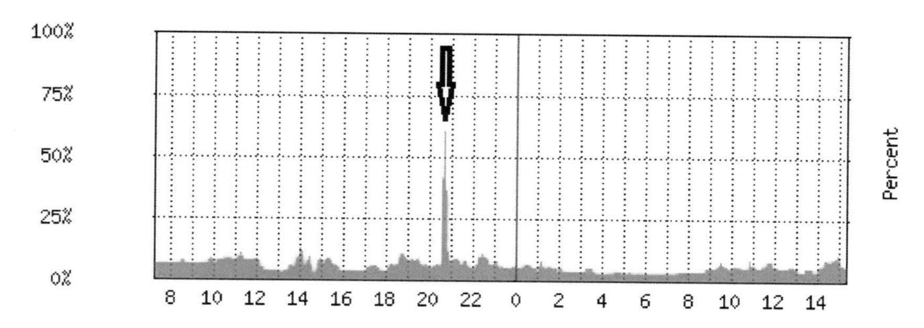

Figure 4.9. MRTG daily CPU usage of router *B* during running our algorithm

Our algorithm is able to detect almost all illegal requests in the form of UDP and TCP regardless of their source port, destination port and IP address.

4.1 Investigation of attack detection in case of timing

Table 4.1 and Table 4.2 show the time which was taken to detect source IP address of attackers and mitigating the DDoS attacks in the whole process containing detection and mitigation as well for different router threshold setting.

Table 4.1. Time taken for the whole detection and mitigation process for router setting of $\beta = 4$ and $\alpha = 0.1$.

# of experimen t	1	2	3	4	5	6	7	8	9	10
Time elapsed for the whole process in second	123	127	121	129	124	127	124	129	128	126

The average time taken in Table 4.1 is 125.8 seconds whereas this average time in Table 4.2 is 117.5 seconds which means that using lower threshold level β will lead to faster detection of the attacks.

Table 4.2. Time taken for the whole detection and mitigation process for router setting of $\beta = 3$ and $\alpha = 0.1$.

# of experiment	1	2	3	4	5	6	7	8	9	10
Time elapsed for the whole process in second	113	120	114	120	116	120	115	121	119	117

4.2 False-positive and False-negative error investigation

A false positive error, commonly called a "false alarm" is a result which indicates a given condition has been fulfilled, when it actually has not been fulfilled. On the other hand, a false negative error is where a test result indicates that a condition was failed, while it actually was successful.

We define the probability (P_{-}) of not detecting the attack traffic (i.e., the probability of the false-negative errors) and the probability (P_{+}) of erroneously detecting an attack (i.e., the probability of false-positive errors), which are calculated from the following equations:

$$P_{-} = \frac{\text{Number of attacks not detected}}{\text{Total number of attacks}}$$

$$P_{+} = \frac{\text{Number of points errorneously detected as attacks}}{\text{Total number of attacks}}$$

We generated fake packets that behave as attacks from 15 different locations in the network by 15 independent Mikrotik devices. Hence, the number of attackers in our experiment is 15 (N_1 *to* N_{15} in figure 3.1) attacks in total.

As shown in Table 4.3 and Table 4.4 the results of repeating the same experiment for 10 rounds in the real network imply on the accuracy and trustworthiness of the proposed algorithm. Note that in Table 4.1, Table 4.3, and Table 4.5 all the measurements are considered regarding to router threshold setting of $\beta = 4$ with an inertia ratio $\alpha = 0.1$. All the measurements in Table 4.2, Table 4.4, and Table 4.6 are for router threshold setting of $\beta = 3$ with an inertia ratio $\alpha = 0.1$.

Table 4.3. Results for False-negative and False-positive error for router setting of $\beta = 4$ and $\alpha = 0.1$

# of experiment	1	2	3	4	5	6	7	8	9	10
# of attacks not detected	1	2	1	0	0	2	1	1	0	2
probability (P_-) of not detecting attack traffic	6.66 %	13.3 %	6.66 %	0 %	0 %	13.3 %	6.66 %	6.66 %	0 %	13.3 %
# of points erroneously detected as attacks	0	1	0	0	0	0	0	1	0	0
# of points detected as attacks in total	14	12	14	15	15	13	14	13	15	13
probability (P_+) of erroneously detecting attack	0%	6.66 %	0%	0 %	0 %	0%	0%	6.66 %	0 %	0%

Table 4.4. Results for False-negative and False-positive error for router setting of $\beta = 3$ and $\alpha = 0.1$

# of experiment	1	2	3	4	5	6	7	8	9	10
# of attacks not detected	0	1	0	0	0	0	0	1	0	0
probability (P_{-}) of not detecting attack traffic	0 %	6.66%	0%	0%	0%	0%	0%	6.66%	0%	0%
# of points erroneously detected as attacks	3	1	4	2	1	4	4	1	3	3
# of points detected as attacks	12	13	11	13	14	11	11	13	12	12
probability (P_{+}) of erroneously detecting attack	25 %	6.66%	36.4%	15.4%	6.6 %	26.6%	26.6%	6.66%	20.0%	20.0%

4.3 Performance Metrics Used

The performance of our detection scheme is evaluated with three following metrics: detection rate, false-positive alarms (P_+), and system overhead. All the metrics are measured under different DDoS attacks using TCP, UDP, and ICMP. The detection rate R_d of DDoS attacks is defined by the following ratio (Y. Chen, K. Hwang, and Wei-Shinn Ku, 2007):

$$R_d = \frac{a}{n}$$

where a is the number of DDoS attacks detected in the simulation experiments, and n is the total number of attacks generated by the Mikrotik routers during each experiment which is 15. Table 4.5 and Table 4.6 show the measured detection ratio over the whole 10 independent experiments.

Table 4.5. Detection rate R_d for $\beta = 4$ and $\alpha = 0.1$

# of experiment	1	2	3	4	5	6	7	8	9	10
Detection ratio (***R***$_d$)	93.3 %	80 %	93.3 %	100 %	100 %	86.6 %	93.3 %	86.6 %	100 %	86.6 %

Table 4.6. Detection rate R_d for $\beta = 3$ and $\alpha = 0.1$

# of experiment	1	2	3	4	5	6	7	8	9	10
Detection ratio (***R_d***)	80%	86.6%	73.3%	86.6%	93.3%	73.3%	73.3%	86.6%	80%	80%

4.4 Trade off

By taking a glance view on the Table 4.1 and Table 4.2 we can see that the number of attacks not detected in Table 4.1 is more than Table 4.2 due to the differences between the two parameters of β and α. The most important thing to achieve the best results is to keep false-positive alarm rate minimum. If we look carefully on the result of these two tables we can see that the number of points erroneously detected as attacks in the Table 4.1 are less than the Table 4.2 by far. Figure 4.10 compares the detection rates R_d between the results achieved by these two tables.

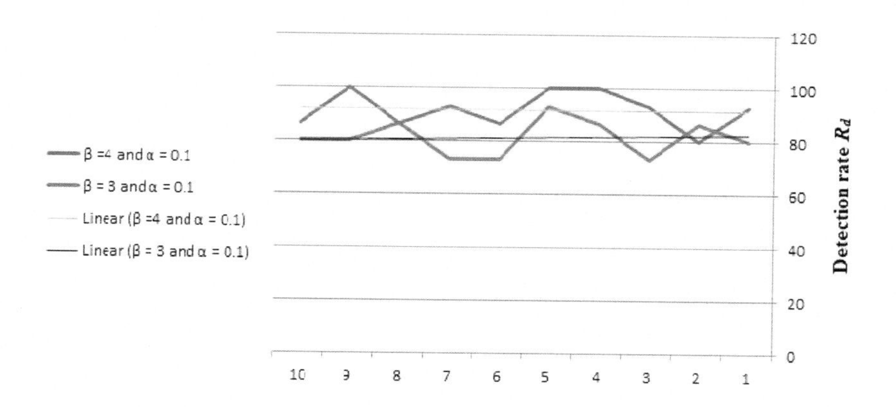

Figure 4.10. Detection ratio R_d with different router threshold levels *(α* and *β).*

The linear lines in Figure 4.10 are trend line for two different threshold levels. These two trend line show a trade off on average detection ratio R_d for different router threshold settings. According to the achieved results we can conclude that using our algorithm with the optimal router threshold setting $\beta \geq 3.5$ with an inertia ratio $\alpha = 0.1$ give us the best results.

4.5 Summary

We must underline the importance of two parameters that plays the most challenging role to select suitable algorithm to identify and mitigate DDoS attacks and they are:

1. Achieve higher detection rate
2. Achieve low detection time

To achieve higher detection rate in the lowest time we also have to consider and keep the two main factors of false-positive and false-negative errors optimum. In this work has been tried to satisfy these factors as well.

CHAPTER 5

CONCLUSIONS AND RECOMMENDATIONS

5.1 Conclusions

To avoid any widespread damage to the victims system, DDoS attacks have to be detected at their early launching stage. We have developed a defence and a distributed detection mechanism to protect servers and websites against distributed denial of service (DDoS) attacks. We proposed a mechanism based on collaboration among the two most important gateways of attacks which are edge routers and victims firewall routers. Such a network is an effective way to enhance the attacks detection rate, provide attack alerts and protect legitimate traffic.

We proposed a new detection method to increase the attacks detection accuracy, which samples the current incoming traffic and CPU usage of the destination target of attack and calculates the difference from average. It then compares it to the predefined parameter β to distinguish whether or not the attacks happened. The threshold β measure the magnitude of traffic surge over the average traffic value or the magnitude of CPU surge over the average CPU value. If the firewall senses any attacks it will send a request to any source edge routers to collaborate with them in order to detect and mitigate attacks.

Based on this result achieved, we introduced a new attack detection method. Through real experiments, we have shown that our method can detect and block attack packets quickly. We have also shown the effects of attacker-side defence and the effectiveness of our method.

5.2 Recommendation

In the large networks for a faster response to detect and mitigate attacks, we recommend the categorization of edge router in j groups. It would be more efficient to send a collaborative Flag request to the whole edge routers simultaneously. This categorization can be done in first step by dividing the number of available edge router to two independent groups and each also divide by two. The number of division iteration would depend on how fast we expect to find attacks and the number of available edge routers.

REFERENCES

A. Hussain, J. Heidemann, and C. Papadopoulos, "*Identification of Repeated Denial of Service Attacks,*" Proc. INFOCOM '06, Apr. 2006.

A. J. P. Jeckmans, "*Practical Client Puzzle from Repeated Squaring*," ed, 2009.

A. Kuzmanovic and E.W. Knightly, "*Low-Rate TCP-Targeted Denial of Service Attacks (The Shrew vs. the Mice and Elephants)*," Proc. ACM SIGCOMM 2003, Aug. 2003.

A. Yaar and D. Song, "*SIFF: A Stateless Internet Flow Filter to Mitigate DDoS Flooding Attacks,*" Proc. 2004 IEEE Symp. Security and Privacy, 2004.

A.D. Keromytis, V. Misra, and D. Rubenstein, "*SOS: An Architecture for Mitigating DDoS Attacks,*" IEEE J. Selected Areas in Comm., vol. 22, no. 1, pp. 176-188, Jan. 2004.

C. Jin, H. Wang, and K.G. Shin, "*Hop-Count Filtering: An Effective Defense against Spoofed Traffic,*" Proc. ACM Conf. Computer and Comm. Security (CCS '03), Oct. 2003.

C. Papadopoulos, R. Lindell, J. Mehringer, A. Hussain, and R. Govindan, "*COSSACK: Coordinated Suppression of Simultaneous Attacks,*" Proc. Third DARPA Information Survivability Conf. and Exposition (DISCEX-III '03), pp. 2-13, 2003.

Cisco IOS Security Configuration Guide, Release 12.2 "*Configuring Unicast Reverse Path Forwarding,*" pp. SC-431-SC-446,

http://www.cisco.com/univercd/cc/td/doc/product/software/ios122/122cgcr/fsecur_c/fothersf/scfrpf.pdf. 2006, 2006.

D. Dean and A. Stubblefield, "*Using client puzzles to protect TLS,*" presented at the Proceedings of the 10th conference on USENIX Security Symposium - Volume 10, Washington, D.C., 2001.

D. J. Bernstein. "*Better price-performance ratios for generalized birthday attacks*," 2007.

D. Moore, G. Voelker, and S. Savage, "*Inferring Internet Denial-of-Service Activity,*" Proc. 10th Usenix Security Symp., 2001.

Drew Dean, Adam Stubblefield, "*Using Client Puzzles to Protect TLS*"

E. H. McKinney (1966) "*Generalized Birthday Problem*," American Mathematical Monthly 73, 385–387.

G. Carl, G. Kesidis, R. Brooks, and S. Rai, "*Denial-of-Service Attack Detection Techniques,*" IEEE Internet Computing, Jan./Feb. 2006.

G. Karame and S. Čapkun, "*Low-Cost client puzzles based on modular exponentiation,*" in Computer Security – ESORICS 2010. vol. 6345, D. Gritzalis, et al., Eds., ed: Springer Berlin / Heidelberg, 2010, pp. 679-697.

H. Aljifri, "*IP Traceback: A New Denial-of-Service Deterrent,*" IEEESecurity and Privacy, pp. 24-31, May/June 2003.

H. Jiang and C. Dovrolis, "*Why Is the Internet Traffic Bursty in Short Time Scales,*" Proc. ACM SIGMETRICS '05, June 2005.

H. Wang, D. Zhang, and K. Shin, "*Change-Point Monitoring for the Detection of DoS Attacks,*" IEEE Trans. Dependable and Secure Computing, vol. 1, Oct.-Dec. 2004.

http://www.mikrotik.com/

J. Ioannidis and S.M. Bellovin, "*Implementing Pushback: Router-Based Defense against DDoS Attacks,*" Proc. Network and Distributed System Security Symp. (NDSS '02), Feb. 2002.

J. Mirkovic and P. Reiher, "*D-WARD: A Source-End Defense against Flooding DoS Attacks,*" IEEE Trans. Dependable and Secure Computing, pp. 216-232, July 2005.

J. Mirkovic, G. Prier, and P. Reiher, "*Attacking DDoS at the Source,*" Proc. 10th IEEE Int'l Conf. Network Protocols, Nov. 2002.

Jie Yu, Chengfang Fang, Liming Lu, *Zhoujun Li, A Lightweight Mechanism to Mitigate Application Layer DDoS Attacks*,

K. Houle et al., "*Trends in Denial of Service Attack Technology,*" www.cert.org/archive/pdf/, 2001.

K. Hwang, M. Cai, Y. Chen, and M. Qin, "*Hybrid Intrusion Detection with Weighted Signature Generation over Anomalous Internet Episodes,*" IEEE Trans. Dependable and Secure Computing, vol. 4, no. 1, pp. 41-55, Jan.-Mar. 2007.

K. Park and H. Lee, "*On the Effectiveness of Route-Based Packet Filtering for Distributed DoS Attack Prevention in Power-Law Internets,*" Proc. ACM SIGCOMM, pp. 15-26, 2001.

Kihong Park and Heejo Lee, "On the effectiveness of route-based packet filtering for distributed DoS attack prevention in power-law internets," in *Proceedings of ACM SIGCOMM*, 2001, pp. 295–306

M. Walfish, M. Vutukuru, H. Balakrishnan, D. Karger, and S. Shenker, "*DDoS Defense by Offense,*" Proc. ACM SIGCOMM '06, Sept. 2006.

Mark Fullmer and Steve Romig. The OSU Flowtools Package and Cisco Netflow logs. In *Proceedings of the 2000 USENIX LISA Conference*, New Orleans, LA, December 2000.

Nicholas A. Fraser, Douglas J. Kelly, Richard A. Raines, Rusty O. Baldwin, and Barry E. Mullins. *"Using Client Puzzles to Mitigate Distributed Denial of Service Attacks in the Tor Anonymous Routing Environment"*, In Proceedings of ICC 2007

P. Ferguson and D. Senie, "*Network Ingress Filtering: Defeating Denial of Service Attacks which Employ IP Source Address Spoofing,*" RFC 2827, 2000.

P. Ning, S. Jajodia, and X.S. Wang, "*Abstraction-Based Intrusion Detection in Distributed Environment,*" ACM Trans. Information and System Security, pp. 407-452, Nov. 2001.

Q. Li, E. Chang and M. Chan. *On the Effectiveness of DDoS Attacks on Statistical Filtering*, In Proceedings of INFOCOM'05, 2005.

Q. Li, E.C. Chang, and M.C. Chan, "*On the Effectiveness of DDoS Attacks on Statistical Filtering,*" Proc. 2005 IEEE INFOCOM, 2005.

R. Blazek et al., "*A Novel Approach to Detection of DoS Attacks via Adaptive Sequential and Batch-Sequential Change-Point Detection Methods,*" Proc. IEEE Workshop Information Assurance and Security, June 2001.

R. C. Merkle, "*Secure communications over insecure channels*," Commun. ACM, vol. 21, pp. 294-299, 1978.

S. Bellovin, J. Schiller, and C. Kaufman, *Security Mechanism for the Internet*, IETF RFC 3631, 2003.

S. Chen and Q. Song, "*Perimeter-Based Defense against High Bandwidth DDoS Attacks,*" IEEE Trans. Parallel and Distributed Systems, vol. 16, no. 6, June 2005.

S. Kandula, D. Katabi, M. Jacob, and A. Berger, "*Botz-4-Sale: Surviving Organized DDoS Attacks That Mimic Flash Crowds,*" Proc. Second Symp. Networked Systems Design and Implementation (NSDI '05), May 2005.

S. Kent and R. Atkinson, *Security Architecture for the Internet Protocol*, IETF RFC 2401, 1998.

S. Ranjan, R. Swaminathan, M. Uysal, and E. Knightly, "*DDoSResilient Scheduling to Counter Application Layer Attacks under Imperfect Detection,*" Proc. INFOCOM '06, Apr. 2006.

Stephen M. Specht and Ruby B. Lee, *Distributed Denial of Service: Taxonomies of Attacks, Tools, and Countermeasures.* Proceedings of the 17th International Conference on Parallel and Distributed Computing Systems, 2004 International Workshop on Security in Parallel and Distributed Systems, pp. 543-550, September 2004

T. Gil and M. Poletto, "*MULTOPS: A Data-Structure for Bandwidth Attack Detection,*" Proc. 10th Usenix Security Symp., Aug. 2001.

T. Peng, C. Leckie, and K. Ramamohanarao, "*Detecting Distributed Denial of Service Attacks by Sharing Distributed Beliefs,*" Proc. Eighth Australasian Conf. Information Security and Privacy (ACISP '03), July 2003.

T. Ryutov, L. Zhou, C. Neuman, T. Leithead, and K.E. Seamons, "*Adaptive Trust Negotiation and Access Control,*" Proc. ACM Symp. Access Control Models and Technologies (SACMAT '05), June 2005.

W.E. Leland, M.S. Taqqu, W. Willinger, and D.V. Wilson, "*On the Self-Similar Nature of Ethernet Traffic,*" Proceedings of the ACM SIGCOMM 1993 Symposium on Communications Architectures, Protocols, and Applications, pp. 183-193, September 1993.

Wikipedia, http://en.wikipedia.org/wiki/Birthday_attack

Wu-chang Feng and Ed. Kaiser, "*mod_kaPoW: mitigating DoS with transparent proof-of-work,*" presented at the Proceedings of the 2007 ACM CoNEXT conference, New York, New York, 2007.

X. Wang, S. Chellappan, P. Boyer, and D. Xuan, "*On the Effectiveness of Secure Overlay Forwarding Systems under Intelligent Distributed DoS Attacks*," IEEE Trans. Parallel and Distributed Systems, vol. 17, no. 7, July 2006.

X. Wang, S. Chellappan, P. Boyer, and D. Xuan, "*On the Effectiveness of Secure Overlay Forwarding Systems under Intelligent Distributed DoS Attacks,*" IEEE Trans. Parallel and Distributed Systems, vol. 17, no. 7, July 2006.

Y. Chen and K. Hwang, "*Collaborative Change Detection of DDoS Attacks on Community and ISP Networks,*" Proc. IEEE Int'l Symp. Collaborative Technologies and Systems (CTS '06), May 2006.

Y. Chen and K. Hwang, "*Collaborative Detection and Filtering of Shrew DDoS Attacks Using Spectral Analysis,*" J. Parallel and Distributed Computing, special issue on security in grids and distributed systems, pp. 1137-1151, Sept. 2006.

Y. Chen, K. Hwang, and Wei-Shinn Ku "*Collaborative Detection of DDoS Attacks over Multiple Network Domains,*" IEEE TRANSACTIONS ON PARALLEL AND DISTRIBUTED SYSTEMS, VOL. 18, NO. 12, DECEMBER 2007.

Y. Gao, *et al.*, "*Efficient trapdoor-based client puzzle against DoS attacks*," in *Network Security*, S. C. H. C. H. Huang, *et al.*, Eds., ed: Springer US, 2010, pp. 229-249.

Y. Kim, J.Y. Jo, and F. Merat, "*Defeating Distributed Denial-of-Service Attack with Deterministic Bit Marking,*" Proc. IEEE GLOBECOM, Dec. 2003.

Y. Kim, J.Y. Jo, H.J. Chao, and F. Merat, "*High-Speed Router Filter for Blocking TCP Flooding under Distributed Denial-of-service Attack,*" Proc. IEEE Int'l Performance, Computing, and Comm. Conf., Apr. 2003.

Y. Kim, W.C. Lau, M.C. Chuah, and H.J. Chao, "*PacketScore: Statistics-Based Overload Control against Distributed Denial of Service Attacks,*" Proc. INFOCOM '04, 2004.

Y. Kim, W.C. Lau, M.C. Chuah, and H.J. Chao, "*PacketScore: Statistics-Based Overload Control against Distributed Denial-of-Service Attacks,*" Proc. IEEE INFOCOM, Mar. 2004.

Y. Xu and R. Gue´rin, "*On the Robustness of Router-Based Denial of-Service (DoS) Defense Systems,*" ACM

SIGCOMM Computer Comm. Rev., vol. 35, no. 3, July 2005.

81956969R00055

Made in the USA
San Bernardino, CA
12 July 2018